Study Guide for
Lamanna and Riedmann's

Marriages and Families
MAKING CHOICES IN A DIVERSE SOCIETY

SIXTH EDITION

David Treybig
Matagora Beach, TX

Wadsworth Publishing Company
I(T)P® An International Thomson Publishing Company

Belmont • Albany • Bonn • Boston • Cincinnati • Detroit • London • Madrid • Melbourne
Mexico City • New York • Paris • San Francisco • Singapore • Tokyo • Toronto • Washington

Printed in the United States of America
2 3 4 5 6 7 8 9 10

For more information, contact Wadsworth Publishing Company.

Wadsworth Publishing Company
10 Davis Drive
Belmont, California 94002, USA

International Thomson Publishing Europe
Berkshire House 168-173
High Holborn
London, WC1V 7AA, England

Thomas Nelson Australia
102 Dodds Street
South Melbourne 3205
Victoria, Australia

Nelson Canada
1120 Birchmount Road
Scarborough, Ontario
Canada M1K 5G4

International Thomson Editores
Campos Eliseos 385, Piso 7
Col. Polanco
11560 México D.F. México

International Thomson Publishing GmbH
Königswinterer Strasse 418
53227 Bonn, Germany

International Thomson Publishing Asia
221 Henderson Road
#05-10 Henderson Building
Singapore 0315

International Thomson Publishing Japan
Hirakawacho Kyowa Building, 3F
2-2-1 Hirakawacho
Chiyoda-ku, Tokyo 102, Japan

ISBN 0-534-50555-4

Contents

PREFACE

This STUDY GUIDE contains two sets of instructions about how to use the STUDY GUIDE. For students who know how to study and how to use a STUDY GUIDE, one set of instructions —the first set—is very brief. For students who want more detailed information about how to study and how to use the STUDY GUIDE, a more detailed set of instructions follows the brief instructions. Feel free to read both sets of instructions if you wish. Most students find the instructions to be quite helpful, reporting a significant grade increase after making the STUDY GUIDE an integral part of their study routine.

BRIEF INSTRUCTIONS

Although there is no foolproof way to study, to learn, and to prepare efficiently to take examinations, the following suggestions may assist you.

1. Before you read a textbook chapter, examine the **Chapter Overview** at the beginning of the chapter and the **Chapter Summary** at the end of the chapter. This will give you some idea of what the chapter is about. Armed with this intellectual road map, you will find the chapter easier to understand and to remember.

2. Read the chapter section by section, pausing between sections in order to *think* about what you have read. Please do not read the entire chapter without pausing, because might have difficulty retaining what you have read. Read a section, and then *stop to think* about what you have read. When you feel comfortable that you know and Studying like this may take more time, but it tends to produce more learning and better grades.

3. Be sure to pay attention to the charts, graphs, case studies, and even the written material accompanying the illustrations and photographs. The instructor may include questions on this material. If the material was not important, it would not be included in the text. *Do not memorize it. Understand it.*

4. After reading the chapter as indicated above (some students like to make notes as they read) reread the Chapter Overview and Chapter Summary in the STUDY GUIDE. Then answer the questions in the STUDY GUIDE. After you have answered them, check the answer using the answer key at the end of each STUDY GUIDE chapter. If you miss a Question, review the material in the text to find out why you missed it.

5. Remember, in a course like this, one of the goals is to understand the material. *Do not memorize it. Understand it.*

6. Finally, pause between studying this material before going on to do something else. Take a few moments to reflect on the material you have studied. This helps to retain what you studied. And remember to allow for a thorough review before taking examinations.

DETAILED INSTRUCTIONS

This **STUDY GUIDE** is intended to help you study **MARRIAGES AND FAMILIES,** 6th Edition, by Lamanna and Riedmann. The **STUDY GUIDE** provides helpful suggestions. It gives a brief overview of each chapter and alerts you to important terms, concepts, theories, and research results. It can even give you some sample examination questions so you can test your understanding of the text. However, the **STUDY GUIDE** is not a substitute for the text. Instead, it complements the text and properly used should help you to understand the text better, retain for a longer time that which you have learned, and also help you make better grades on exams. Ultimately, each student must find her or his own ways of effective study. But the suggestions below have proved very effective for many students.

FORMAT OF THE STUDY GUIDE

The study units in the **STUDY GUIDE** follow the format and organization of the text, **MARRIAGES AND FAMILIES,** 6th edition, by Lamanna and Riedmann. Just as in the text, the **STUDY GUIDE** contains a study unit for each of the chapters in the text and for the appendices. Each study unit contains the following:

1. A **Chapter Overview** that briefly notes the topics to be covered in the chapter.

2. A **Chapter Summary** that captures the essential points in a few paragraphs. Because of the superior quality of the text's chapter summaries, the Chapter Summaries are based primarily on the text's summaries.

3. **Key Terms** that should be fully understood. The student should fully understand each key term and be able to define it using his or her own words as well as using technical sociological terms. The student should be able to give one or more examples of each key term and explain *why* the example is an example of the key term.

4. **Completion Items**, or sentences in which the student writes in the appropriate key term from the list of Key Terms given previously in each chapter.

5. **Key Research Studies**. It is critical that students understand the findings of important research. The Key Research Studies alert students to the research to which they should pay especially close attention in the text. Though students should be alert to all research findings discussed in the text, the Key Research Studies should be closely examined regarding the purpose of the research, how the research was conducted, what was found, and what the authors think the research means in terms of marriages and families.

   Typically, the names of the researchers are given along with their research. Always, however, the most important thing to be learned is what the researchers *found*, not the names of the researchers. If you can recall the names of the researchers, that is admirable and of course desirable. But the essential thing to be learned is what the researchers discovered, found, or concluded as a result of their efforts. The sample test questions will help you to develop this important study skill.

6. **Key Theories.** The main purpose of theories is to help us to understand *why* things are as they are. Though there may be other theories in the text chapters, the Key Theories are central to understanding the chapter material. Therefore, students should be certain that they have a complete understanding of the Key Theories. Here, too, the most important things to learn are the names of the Key Theories and the meanings of the Key Theories, not the name of the persons who proposed the theories. Again, it is admirable and desirable if you can remember the theorists' names, but the essential thing is to remember what the theory is called and to understand what it means —*how it explains* what it is trying to explain.

7. **True-False Questions.** These questions are included in the **STUDY GUIDE** to help you test your knowledge of the text material and because some instructors include such questions on examinations. The **True-False** questions are not intended to be "tricky." They are written to be *obviously* true or *obviously* false.

8. **Multiple-Choice Questions**. These questions are included for the same reasons the true-false questions are included—to allow you to test your knowledge of the chapter material and because some instructors use multiple-choice questions on examinations.

9. **Short-Answer Essay Questions** and **Essay Questions** are included in this **STUDY GUIDE**. The student should be able to answer both types of essay questions because instructors may use these questions like them when constructing an examination. Also, answering these questions can be an excellent way to review the chapter material. Many students write out the answers to these questions and then review their written answers before taking examinations. It can be a very effective study and review technique.

10. **Answers to completion items, true-false-questions, and multiple-choice questions.** If you miss questions, you may not fully understand the chapter materials. Likewise, to the extent that you are correct in answering these questions, you probably have a correspondingly high understanding of the textbook material. Although you will probably not get *exactly* the same questions, word for word, on your in-class examinations as you find in this **STUDY GUIDE**, you will probably find a strong similarity regarding question type, wording, and topic coverage or content.

We recommend, therefore, that you do not study the **STUDY GUIDE** *questions*, but instead study the Lamanna and Reidmann *text*. Use the **STUDY GUIDE** to *help* study the text, to assess how well you *understand* the text, and as a *review* technique before examinations.

This STUDY GUIDE can help you to learn more effectively, retain more information, understand more, and get better grades on examinations. But don't assume that these questions are the only questions that will be on the examinations or tests you take in class. There will be other questions on the examinations or tests, but they will be on a similar question type, wording, and topic coverage or content.

If you merely try to memorize the answers to questions in this STUDY GUIDE, you may do poorly on in-class examinations. But if you use the STUDY GUIDE to help you study the text, to test your knowledge of the text once you have studied it, and as a review technique before examinations, you should expect to do well on in-class examinations.

Study the text as suggested in the STUDY GUIDE. Then use the STUDY GUIDE questions as a rough indicator of how effective your studying has been.

SUGGESTIONS FOR EFFECTIVE STUDY SKILLS:
Studying Smarter, Not Harder

How sure are you that your study skills are good ones? Are you sure that they are the best they can possibly be? Can they be improved? It is now recognized that the great majority of students have poor study skills. Most students who study but still get low grades get them because their study skills are low. If you spend considerable time studying but don't do as well on examinations as you would like, then your study skills probably aren't as good as they should be. But they can be improved! You might consider following the suggestions below as ways to improve the effectiveness of your study time.

Amount of Time Spent Studying

How much time should be spent studying? It differs from person to person and from course to course, but a useful rule of thumb is that a persons should spend from one to three hours studying for every hour spent in class. Most students seriously underestimate the amount of time that needs to be spent studying in order to do well in a course. Put different — and bluntly! — most students want to get good grades but do not spend *enough* time studying to get the grades they want.

The text does not need to be merely read. It needs to be *studied.* And studying it is very different from just reading a chapter from beginning to end without pause. To study the text effectively takes time. Likewise, taking class notes does little good if they are never studied. And merely reading through class notes is not equivalent to studying them.

Realistically appraise your personal schedule and set aside sufficient study time for each course you are in. Decide which days and times you will study, write those study sessions on your daily calendar, and then stick with it. Self-discipline and assertiveness are needed in order to keep to your study schedule.

And make no mistake about it, a schedule that is followed is necessary if you are to get the kinds of grades and levels of learning that you want. If friends or kin knock on your door while you are studying and invite you to do something else, just tell them "No," and suggest a later time when you expect to be available. If your studies are important, establish a definite schedule and stick with it!

Studying a Text: Some Strategies That Work

Many students who have used the study skills included in this book and that discussed below ordinarily increase their exam grades by one or two letter-grades. If you want your grades to improve, you can probably improve them. It is primarily a matter of knowing how to study and then putting that knowledge into action. Merely reading through a text chapter from beginning to end is not an effective way to learn, understand, and retain the material and get better grades. There are specific strategies for effectively studying textbooks, and if you want to learn them and apply yourself to the task, you can get better grades.

These skills have appeared in various combinations and have been given various names (SQ3R or SQ4R, for example). But regardless of what they are called, the basic skills are the same. If you learn them and use them, you will probably see a significant increase in your level of learning and in your grades. Getting better grades is not so much a matter of having high intelligence as it is a matter of learning and applying easy-to-learn methods of studying.

STUDY GUIDE Exercises and Exam Questions

If you can answer the **STUDY GUIDE** questions as well as the Study Questions at the end of each textbook chapter, you ought to do very well on examinations. There are several reasons for this:

1. The **STUDY GUIDE** questions probably will be identical to the question you get on an examination, but they will be similar to many of the examinations questions in terms of question type, wording, content, and specificity or detail.

2. Think of the **STUDY GUIDE** questions as a sample of the questions you may have on examinations. If you do well on the **STUDY GUIDE** questions, then you should probably do equally well on the examination. ON the other hand, if you do not do very well on the **STUDY GUIDE** questions, then you may not do very well on the examinations unless you increase your level of learning and understanding.

3. The **STUDY GUIDE** can alert you to specific topics about which you should be familiar. If you miss a question, you can "repair the damage" by re-studying or reviewing the text material.

Testing Your Knowledge

The STUDY GUIDE can be used in several ways to test your knowledge of the material in the text. If you use all of them, the results should be increased insight about how well you know the text material, reinforcement of what you learned when you studied the text, and additional learning that occurs as you use proceed through the STUDY GUIDE process, completing section after section of the guide for each textbook chapter.

Key Terms

Begin testing your knowledge with the key terms listed for each chapter in the STUDY GUIDE. Turn each **key term** into a question—and then try to answer it. If the **key term** is "life spiral model," as yourself "What is the life spiral model?" and try to answer the question. If you can answer the question and are sure that your answer is correct, good for you! But if you cannot answer the question or if you feel unsure of your answer, review the chapter material pertaining to that question. Continue in this manner until you have turned all of the **key terms** into questions and, it is hoped, into correct answers. (Incidentally, some students find it very helpful to explain the answers to such questions to another person, preferably to another student in the class. If you just think the answer through, you can sometimes fool yourself into thinking that you know the answer when you don't. But if you have to explain the answer to another person, it becomes clear whether or not you know the material. You can use this technique on other types of questions as well!)

Completion

Next, answer the **completion** items. Read each sentence and fill in the blank with the correct term from the Key Terms section. Read each sentence, scan the list of Key Terms, and write what you takes to be the correct answer in the blank space. When you have done this for all of the **completion** items, check your answers by referring to the answers at the end of each STUDY GUIDE chapter.

If you missed one or more of the **completion** items, review the pertinent material in the textbook chapter. Later, when you are reviewing for an examination, it can help to review your completed answers in this section of the STUDY GUIDE.

Key Research Studies and Key Theories

For the **Key Research Studies** and **Key Theories** sections, turn each research study and each theory into a question that you ask yourself and then try to answer. If the **Key Research Studies** section lists Geerson's research on women's role choices, ask yourself what it was that Geerson found when this research was conducted. If you studied the text and understood it, you should know.

If the **Key Theories** section lists ecological theory of family, ask yourself what ecological theory of family is, what questions it tries to answer, whether it has any particular strengths and weaknesses, and so on. Again, whether you tell the answers to such questions to another (as is recommended) or just think the answer through, or write the correct answer, be sure to review the material if you don't know the answer or if you feel unsure of the answer you have written.

True-False

The **True-False** questions are written so that they are *obviously* true or *obviously* false. None of the **True-False** questions are constructed to be intentionally tricky. Take each question just as it is . Do not add or subtract anything from the question "in your own mind." Is the statement true, according to the text? Or is it false?

Print **T** or **F** in the blank space to the left of each true-false item. When you have completed all of the items, check your answers with the true-false answers section at the end of each chapter in the Study Guide.

If you miss one or more of the true-false items, find the pertinent material in the text and then find out *why* the answer is other than you thought. If you miss a question, be sure to print the correct answer in the Study Guide so you can use these answered questions as part of your review process. (When reviewing, cover the correct answers with a sheet of paper or card, and you can speedily check your response to each question as you review.)

Multiple Choice

When you answer a multiple choice item, do not read the entire question without pause. Instead, read the first part of the item (the "question" part) to locate what the item is about. Before continuing through the alternative answers, stop for a moment and think about that topic. You might want to think briefly about what the correct alternative answer *should* be. Then read the alternative answers. If the item is well-constructed and if you know the text material, your task is easy: Select the correct answer from the alternatives (it should be obvious) and go to the next question.

Well-written multiple-choice items can be an excellent test of your knowledge of the material. In a well-constructed item, the correct alternative answer should not be obvious or stand out because it is "different" from the other alternatives. The student should not be able to infer the answer using only logic. To a student who does not know the text material, *all* of the alternatives should seem correct, even though only one *is* correct. To a student who does understand the text material, the correct alternative answer should seem obviously so, and the other alternatives should seem obviously incorrect.

Print the correct alternative answers in the spaces to the left of the multiple-choice questions. After you have answered all the questions, check your answers with those in the answer section.

Your percent of answers that are correct is a fairly accurate estimate of your level of knowledge of the text material. If you missed a question, return to the text chapter and read the material so that when you review the multiple-choice questions you will answer them (and questions similar to them on the examination) correctly.

Short-answer Essay

You should be able to write a brief but complete short-answer essay (or brief explanation) answer in the space provided in this **STUDY GUIDE.** Some students think that it is easy to write a brief but complete explanation or short-answer essay. They are mistaken. To be able to write a brief but complete answer requires a high level of familiarity with, understanding of, and insight into the assigned material. Of course you ought to be able to write a more complete answer when appropriate or required. But if you find yourself struggling to write a brief explanation that you know is at the same time complete, then you probably have not yet fully mastered the assigned material. You may need to return to the textbook and study the material again until you not only know the details of the topic but also understand the essential core ideas.

Essay

Your instructor may ask you to answer one or more questions in essay form. Instructors differ regarding the length they expect essays to be, but most expect from 1-5 pages. (An essay answer of one-quarter to one-half page is almost surely not an adequate essay answer.) If you are in doubt about the length of the essays expected by your instructor, ask the instructor for clarification.

In addition to answering the practice essay questions provided here, answering the Study Questions at the end of each chapter in the text as if they were essay questions is an excellent way to practice writing essay answers over text material. If you understand the chapter, you should be able to answer these questions or others like them. Answering them will help you to see the overall issues and questions being addressed by the chapter.

CONCLUSION

A study guide cannot work miracles, especially if it is not used properly. But this Study Guide, properly used, should help the student to learn more and do better on examinations. If the Study Guide is used correctly and its suggestions are followed, the student should expect to increase his or her knowledge of **MARRIAGES AND FAMILIES**, and do well on examinations covering text material.

Good luck with your academic efforts, and remember:

> It's *not* a primarily a matter of luck; it's a matter of *studying* wisely and effectively.
>
> Give yourself *enough time* to study.
>
> Study because you *want* to, not because you *have* to.
>
> *Be interested* in the material or find a way to *become* interested in it. One way to become interested in the material is to try to apply it to yourself and to your daily life or to the lives of those you know.
>
> *Don't memorize* the material. *Understand* it!

ACKNOWLEDGMENTS

I thank Dierdre McGill, Assistant Editor at WADSWORTH for asking me to revise the Study Guide for MARRIAGES AND FAMILIES, 6th edition, by Lamanna and Riedmann.

Judie C. Gall provided encouragement and valuable suggestions born from years of experience as a highly skilled professional writer and editor. Shawn C. Carroll gave helpful suggestions that can come only from talented, insightful student users of the **STUDY GUIDE.** Linda T., David P.W., and Robert W. provided socio-emotional and other kinds of support, as did my friend Bill W.

Joyce Kay McLean provided expert assistance in surmounting the considerable technical difficulties inherent in transforming a variety of original electronic materials into a coherent, consistent final product. Without her assistance, the timely completion of this project would have been impossible.

David Treybig, Ph.D.
Matagorda Beach, Texas

CHAPTER 1

FAMILY COMMITMENTS IN A CHANGING SOCIETY

This chapter introduced the subject matter for the course and presented the themes that developed by this textbook. Change and development continue throughout adult life. Marriages and familes are composed of separate, unique individuals who make choices as they are influenced by various factors. All persons in a society, though they are unique as individuals, share some of the same things. Individual and family life consists of a shifting balance of individual freedom and social control. Because individuals and their needs change, marriages and families also inevitably change in response to those shifting needs.

CHAPTER SUMMARY

This chapter began by addressing the challenge of defining the term **family**. The family, because of its relatively small size, face-to-face relationships, tendency to involve the whole person, and the intimate relationships between members, is a _______________ **group**, not a secondary group. It is important to be able to define family because so many social resources are distributed based on family membership, on the basis of who is and who is not a member of a family. Several definitions of the family are presented and discussed, including the **nuclear family, modern family** and ________ - **modern** family.

People make choices, and they are influenced by a number of factors, including ____ **expectations, race and ethnicity, religion, social ________, gender**, and ________________ ________________.
People must make choices and decisions throughout their life courses. Sometimes people are involved in **choosing knowledgeably** by being actively involved in assessing their options, their preferences. And sometimes people simply let things happen to them, enduring with varying levels of success whatever comes their way, in which case they are **choosing ____________________.** When people choose, their choices and decisions simultaneously are limited by social structure and are causes for change in that structure.

Marriages and families are composed of separate, unique individuals. That uniqueness stems partly from the fact that human beings are able to make choices. They have creativity and free will: Nothing they think or do is totally "programmed."

At the same time, all of the individuals in a particular society share some things. They speak the same language and have some common attitudes about work, education, marriages and families. Moreover, within a socially diverse society such as ours, many individuals are part of a racial, ethnic, or religious community or social class that has a distinctive family heritage.

Our culture values both **individualistic (self-fulfillment)** values and **family values** (___________). Whether individualism has gone too far and led to an alarming family decline is a matter of debate.

Even though families fill the important function of providing members a place to belong, finding personal freedom within families is an ongoing, negotiated process.

Adults change, and because they do, marriages and families are not static. Every time one individual in a relationship changes, the relationship itself changes, the result of partners attempting to find ways to alter their relationships to meet their changing needs. Change, then, is one of the main perspectives in this textbook.

Point to Ponder: When you decided to enroll in this course, which of these best described your decision making: choosing knowledgeably, or choosing by default?
(Helpful suggestion: On these Points to Ponder boxed inserts, think your answer through carefully and completely. Many students find it helpful to explain their answer verbally to another student or to write the answers, because doing so helps to think through the material and become clearer about it in your own mind.)

KEY TERMS

You should be able to explain the concepts listed below. In your explanation, try to avoid using the concept you are explaining. You should be able to *give several examples* of each concept and to *explain why* each example is an example.

family 4, 9
primary group 4
nuclear family 5
modern family 5
postmodern family 5
age expectations 15
choosing by default 16
choosing knowledgeably 16
life spiral 17
family values (familism) 18
archival family function 20
individualistic (self-fulfillment) values 20

COMPLETION

Complete the following sentences by selecting the correct alternative from the Key Terms listed on the previous page. Some key terms may be used more than once. Some may not be used at all. Filling in a blank may require more than one word.

1. Andrea wants to become a college freshman now that her children have graduated from college. Her friends tell her she ought to prepare for grandchildren, not for college classes. She knows that her friends think she is too old to go to college. This illustrates what the textbook calls ______________________________.

2. A student secretly worries about whether to not going to college is the right thing to do at this point in life. While the student considers this issue, time goes by, classes and exams are missed, grades decline, and eventually the student is suspended from college. This is an example of what the text calls ______________________________.

3. Tom pondered for two weeks whether or not to work part-time while being a full-time student. He considered the positive and negative aspects of his options, and then made a decision, fully prepared to live with the consequences. This illustrates what the text calls ______________ ______________________.

4. Sara Lee often surprises her friends by her choices, many of which are nontraditional or innovative. Her life, conventional in some ways, takes interesting and unexpected turns, the results of choices she makes. This illustrates the concept of ______________________________.

5. When couples value spending time together in shared activities, perhaps value having or adopting children, value turn to this intimate group as a source of many satisfactions, then family values or ______________________ is expressed.

6. Having year family reunions at which attendance is stressed, acquaintances are renewed, happy memories are relived, and photographs are shared, is an example of the ______________________.

7. Frank considers marrying his long-time friend, but he decides not to because he want to have a lot of time to himself, enjoy the things he wants to enjoy, likes his privacy, and wants to feel free to pursue what tempts him at the moment, with only himself to please. Clearly, Frank is a strong supporter of ______________________________.

8. When each member of a couple tries to go his/her own way, doing innovative and creative things that are not exactly what society expects, and at the same time sustains commitment and time for their relationship, this is an example of complexities, problems, and satisfactions of what the text calls the ______________________________________.

KEY RESEARCH STUDIES

You should be familiar with the main question being investigated and the research findings for the studies listed below. For each, what question were the researchers trying to answer and what was found when the research results were examined? You should understand the question being asked, know what the researchers found, and be able to answer both general and specific questions about the material.

Some facts about families in the United States today (Box 1.1)
Weigert and Hastings: archival family functions
Family ties and immigration (Box 1.2)

KEY THEORETICAL PERSPECTIVES

The purpose of any theory is to help us to see the world more clearly and explain *why* things are they are, thus leading us to greater understanding. You should be familiar with the following key theory, be able to explain it briefly or at length, give examples of it, and answer questions about it.

Etzkowitz and Stein: the life spiral model

Quotation to assess: Read the following quotation and then think about it, assessing or evaluating it in terms of the text material in this chapter. Do not worry about being right or wrong. Just read the quotation and then think about in terms of the concepts, theories, ideas, and research results covered in this chapter of the textbook.

"Biological *possibility* and desire are not the same as biological *need.* Women have childbearing equipment. For them to choose not to use the equipment is no more blocking what is instinctive than it is for a man who, muscles or no, chooses not to be a weightlifter." Betty Rollin (b. 1936), U.S. journalist, author, "Motherhood: Who Needs it?," in ***Look*** (New York, 16 May 1971).

TRUE-FALSE

If you have studied and understand the textbook, you should be able to answer correctly the following true-false questions. Some of the questions are general and some are specific. Each is either obviously true or obviously false. None of the questions are tricky. Answer all of the questions by printing a T or F in the answer space for each question, and then check your answers with the correct answers at the end of this chapter. If you find that you miss questions, review the textbook material to discover *why* your answer is incorrect

_____ 1. Americans these days are both hopeful and apprehensive about marriages and families.

_____ 2. A "primary group" is any large, important group to which a person may belong.

_____ 3. In contrast to the traditional family, the nuclear family is the family of the future.

_____ 4. More and more Americans are living alone.

_____ 5. In 1993 the average size of a U.S. household was 3.8 persons.

_____ 6. People have been postponing marriage in recent years.

_____ 7. The total fertility rate (which represents the number of children born to a hypothetical typical woman during her childbearing years) has generally declined in the U.S.

_____ 8. In 1990, the total fertility rate in the U.S. was 2.9, but a large part of that rate reflects dramatic variations by geographic region.

_____ 9. Young adults are increasingly likely to be living with their parents, a trend reflecting increased economic difficulties, delayed marriage, and high divorce rates.

_____ 10. The divorce rate has dropped slightly since 1981.

_____ 11. In the past 25 years, poverty rates have increased for all American racial/ethnic groups, including nonHispanic whites.

_____ 12. In 1993, approximately 19 million Americans lived below the poverty line of $14,763 for a family of four.

_____ 13. The definition of family adopted by the authors of the textbook includes the requirement that members of a family are committed to maintaining that group over time.

_____ 14. Peter is considering enrolling in a course about couples, families, and singles. But before he does, he finds out whether he needs the class as well as wants it, whether it is highly rated by students who previously took the course, and examines the textbook in the bookstore to review the topics covered in the book. This is an example of knowledgeable decision making.

_____ 15. Social factors limit people's options. An example of this is the fact that American society has never offered the options of legal polygamy.

_____ 16. Federal immigration reforms in 1965 change the criteria for obtaining an immigrant visa, giving high priority to people with needed job skills and low priority to persons with family members already living here.

_____ 17. About 18 percent of white children under age 18 do not live in a two-parent household.

_____ 18. Personally clarifying your own values is an important part of choosing knowledgeably.

_____ 19. According to the text, creating family boundaries is mildly undesirable and/or deviant.

_____ 20. In families, "familism" is normal and healthy, while "individualism" is less so.

_____ 21. The life spiral model assumes that family members have become less predictable than in the past.

MULTIPLE CHOICE

If you have studied and understand the textbook, you should be able to answer correctly the following multiple choice questions. Some of the questions are general and some are specific. Print the correct alternative letter-answer in the space to the left of each question. Check your answers with the correct answers at the end of this chapter. If you understand the chapter very well, you should miss few or none of these questions. For each question you miss, review the pertinent textbook material to understand *why* your answer is incorrect.

_____ 1. The textbook specifically mentions the _________ as something that for many people calls into question the stability of marriage.
   a. the child abuse rate
   b. the cohabitation rate
   c. the percent of the population that is voluntarily single
   d. the divorce rate

_____ 2. The authors of the textbook take the view that _______ are not necessarily bound by legal marriage or by blood or adoption.

a. extended kin
b. traditionally defined nuclear families
c. family members
d. kin defined by consanguinity

_____ 3. Unmarried-couple households increased by more than ___ times in the past 25 years.

a. two
b. three
c. five
d. six

_____ 4. According to one prediction ___ percent of white children and ___ percent of African-American children born in 1980 will spend some time living in single-parent families by the time they reach age 17.

a. 70: 94
b. 62; 75
c. 55; 75
d. none of the above

_____ 5. Which of the following is *not* part of the definition of "family" adopted by the authors of the textbook?

a. the persons form an economic unit
b. the persons consider their identity to be significantly attached to the group
c. the persons are united by a formal or informal ritual that is socially binding
d. the persons are committed to maintaining their family group over time

_____ 6. Ultimately, according to the textbook, the question "What is a family?":

a. is best answered according to shared, local community standards
b. is inherently a practical question that should be answered by such bodies as taxing agencies, lending institutions such as banks, and insurance companies
c. is at its core a religious question, not a political issue
d. has no one correct answer

_____ 7. Which of the following is an example of a "primary group"?

a. a cohabiting couple who have lived together for three years
b. Andrea, LaPrelle, and Ned, who are siblings
c. three divorced men who work in the same office, have lunch together, "work out" at a health club two nights a week, and have breakfast together Saturday mornings
d. all of the above

_____ 8. Marriages among which of the following seem less egalitarian, according to the textbook?
  a. recent immigrants to the U.S.
  b. marriages in which one of the partners is physically disabled
  c. marriages among those age 18-21 and also among those over age 64
  d. residents of the southeastern states

_____ 9. The distribution of wealth in the U.S. has changed:
  a. little since World War II
  b. little since 1935-1937
  c. moderately since Congress changed trade barriers with other nations
  d. dramatically since Congress changed trade barriers with other nations

_____ 10. Which of the following is a recent immigration trend discussed by the textbook?
  a. organized crime profits by selling illegal immigration papers to those desperate to join family members already in the U.S.
  b. many new immigrants to the U.S. save money and send it to needy relatives in their country of origin
  c. approximately 45% of legal immigrants to the U.S. lack skills needed by the occupational structure and thus experience high rates of prolonged joblessness
  d. approximately 30% of legal immigrants to the U.S. become disenchanted and seek to return to their nation of origin

_____ 11. Which of the following illustrates a recent immigration trend discussed by the textbook?
  a. Jorge, Esclarmonde, and Thais are skilled cryo-biologists, and their relatively rare skills facilitate their immigration to the U.S.
  b. Ignacio and Ute-Brigetta marry American visitors to their countries and thus acquire American citizenship
  c. Kevin and Nadine from Gt. Britain have high levels of education, and for this reason find immigration to the U.S. to be both swift and easy
  d. Mukasa and Imkete discover that political persecution in their homeland facilitates immigration to the U.S.

_____ 12. The text cites research finding that working class families tend to value _____ in children, while middle class parents tend to value _____.
  a. promptness and self-determination; education and flexibility
  b. obedience and conformity; self-direction and initiative
  c. self-discipline and goal-setting; self-determination and innovation
  d. orderly work patterns and loyalty to employers; creativity and social skills

_____ 13. Which of the following is an example of an "age expectation" ?
   a. Pete worries about postponing marriage; he feels he should be married by now.
   b. Things can't happen soon enough for Angelica, age 16; if she wants something, immediately is not soon enough.
   c. Angela and Paolo are both age 72 and are retired; they do not expect their days to be filled with hectic demands on their time and in this they are seldom disappointed
   d. all of the above

_____ 14. The textbook warns the reader that _____ have not been widely researched, and there is the risk of wrongly stereotyping if we come to many conclusions about them.
   a. family differences based on religious affiliation
   b. racial family difference
   c. regional family differences
   d. family differences based on partners' political affiliation

_____ 15. Which of the following is not one of the components of choosing knowledgeably?
   a. taking a relatively long time to consider alternatives
   b. rechecking one's decisions
   c. clarifying one's values
   d. none of the above; all are among the components of choosing knowledgeably listed in the textbook

_____ 16. Which of the following is an example of familism?
   a. It is not unusual for families having annual family reunions to find that 60-86 percent of those receiving invitations attend the reunion..
   b. Employment forms in New Zealand, Northern Ireland, and Egypt routinely ask about job applicants' marital status.
   c. A maritime tradition urges the policy of "Women and children first!" should it become necessary to abandon ship during an emergency at sea.
   d. Primatologists refer to monkey "families"; marine biologists use some family-related terms to refer to the status and relationships of whales.

_____ 17. Which of these is the best example of "archival family function"?
   a. In some cultures, human remains are returned to ancestral homelands for burial.
   b. Uncle Dave, the oldest living member of a family, is allowed to retain power and authority because he is the oldest male member of the family.
   c. Tom, being retired, assists his kin by serving as a family librarian and repository for family photographs, letters, and documents.
   d. The Social Security Administration and Internal Revenue Service maintains detailed records about kinship status and economic assistance between family members.

_____ 18. Family life:

a. requires some type of incest taboo to prevent disruptive deviant behavior
b. does not necessarily and inevitably require a power structure
c. thrives only at the expense of the individual person
d. has both costs and benefits

_____ 19. The life spiral includes which of the following concepts:

a. slow but inevitable decline during the course of the family life cycle
b. repeated patterns of crisis and adaptation
c. adaptation to choices and changes two individuals make to each other
d. spending patterns in which family expenditures exceed family income

Your Opinion, Please: Some people feel that it is unfair to a partner to enter a relationship with one set of understandings and agreements and then later — in five or ten years — to change one's mind and want to go in a new direction. What is to be done when a person wants to change but is in a relationship based on previous agreements and assumptions?

SHORT-ANSWER ESSAY QUESTIONS

The following are sample short-answer essay questions — questions of the type you may be asked if your instructor uses questions like these. Even if your instructor does not use questions like these, you can help organize and consolidate your learning if you can answer these questions in a well-organized and complete manner.

Do not think that it is easier to write adequate answer for brief essays than for longer essays. Short-answer essays may be more challenging because answers are required to be both brief and complete. And after you have answered these questions below, construct some similar questions of your own devising, and then answer them. The Study Questions at the end of each textbook chapter make very good short-answer essay questions.

1. How can choice be *both* pressure and freedom?

2. How do age expectations affect choices people make regarding their personal relationships?

3. Explain clearly the life spiral concept and give an example of it.

ESSAY

The following are sample essay questions — questions of the type you may be asked if your instructor uses essay questions. Even if your instructor does not use essay questions, you can help organize and consolidate your learning if you can answer these questions in a well-organized and complete manner. Of these essay questions, the third essay question is usually the most challenging.

1. According to the text, in what ways are personal choices affected by social influences such as historical events, race and ethnicity, social class, and age expectations? Be sure to support your answer with data or facts from the text.

2. In a well-organized essay, explore the idea of change as a usual rather than as an unusual aspect of family life. Support your answer with material from the text.

3. Karl Marx said that it is true that people make history, but they do not do it independently or under conditions of their own choosing. To what extent is this true about the history of individuals, couples, and families?

ANSWERS

CHAPTER SUMMARY

a **primary group**
and **post-modern** family
religion, **social class**
gender and **historical events**
choosing **by default**
family values **(familism)**

COMPLETION

1. age expectations
2. choosing by default
3. choosing knowledgeably
4. life spiral
5. familism
6. archival family function
7. individualistic (self-fulfillment) values
8. life spiral

TRUE-FALSE

1. T
2. F
3. F
4. T
5. F
6. T
7. T
8. F
9. T
10. T
11. T
12. F
13. T
14. T
15. T
16. F
17. T
18. T
19. F
20. F
21. T

MULTIPLE CHOICE

1. D
2. C
3. D
4. A
5. C
6. D
7. D
8. A
9. A
10. B
11. A
12. B
13. A
14. C
15. A
16. A
17. C
18. D
19. C

CHAPTER 2

EXPLORING THE FAMILY

The purpose of theoretical perspectives — sometimes referred to as "theories" — is to give us a set of ideas that help us to understand *why* things are as they are. Why do all societies have some type of family institution or system? Theoretical perspectives help understand why. There are several theoretical perspectives about family, each concentrating on different aspects of the realities of family living.

The purpose of research is not only to provide accurate, complete information, but also to test various theories to discover which ones are most explanatory and most consistent with that which the facts discovered by research. Just as there are several theoretical perspectives about family, so are there a number of research methods that can be used by those interested in studying families.

CHAPTER SUMMARY

Different **theoretical perspectives** illuminate or explain various features of families. This should not be surprising. After all, no one theory can explain everything. Instead, theoretical perspectives tend to focus on specific features or aspects of families. **Family ecology, family development, structure-**___________________, **institutional, interaction, exchange, family systems**, and the **conflict/**__________________ perspectives have in common that each attempts to explain *why* families are as they are. The perspectives differ in the specific aspects or features they attempt to explain. Of exchange theory and family ecology theory, we should not think that one is correct and the other incorrect. It may be more helpful to observe that the two theories do not attempt to explain the same aspects of family living.

The **family ecology perspective** focuses on how families are affected or influenced by the ___________________________ which surround them. As illustrated by government's formation of family policy, families are both influenced by their environment and are capable of affecting or changing that environment in ways that can either improve or worsen the quality of family life. A strength of the family ecology perspective is that it sensitizes us to issues that are not addressed in other theories, but an inherent weakness is that this perspective is so ___________________________ that it includes almost everything and, by doing so, loses some of its explanatory power.

The **family development perspective** concentrates on how families ______________________ over time, sensitizing us to important family transitions and challenges. But its usefulness is somewhat reduced by this theory's assumption that families have a common trajectory or life course.

A view of the family as a social institution whose values, norms, and activities are directed toward the performance of certain functions for society and for its members, is the perspective of

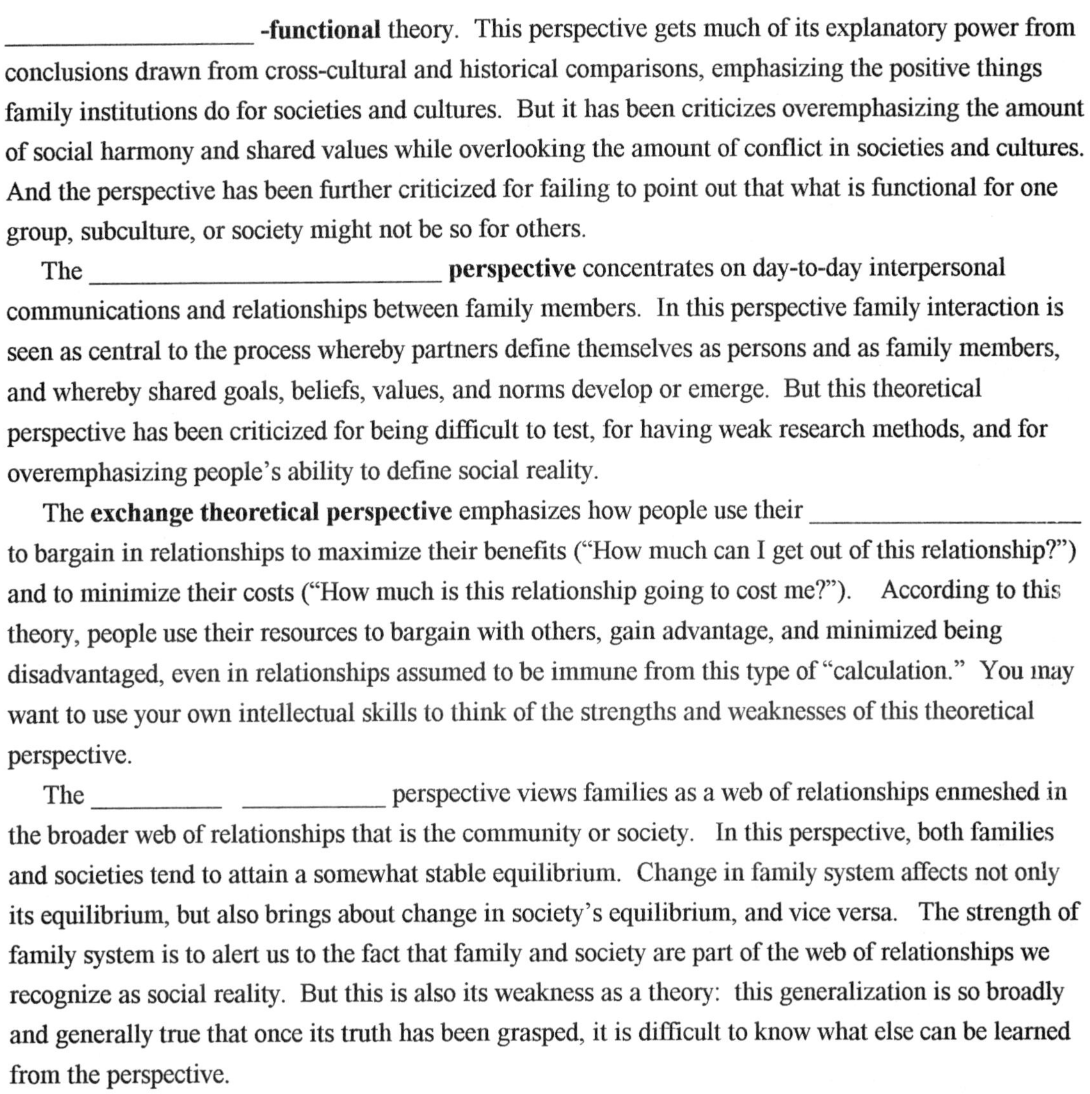

__________________ **-functional** theory. This perspective gets much of its explanatory power from conclusions drawn from cross-cultural and historical comparisons, emphasizing the positive things family institutions do for societies and cultures. But it has been criticizes overemphasizing the amount of social harmony and shared values while overlooking the amount of conflict in societies and cultures. And the perspective has been further criticized for failing to point out that what is functional for one group, subculture, or society might not be so for others.

The __________________________ **perspective** concentrates on day-to-day interpersonal communications and relationships between family members. In this perspective family interaction is seen as central to the process whereby partners define themselves as persons and as family members, and whereby shared goals, beliefs, values, and norms develop or emerge. But this theoretical perspective has been criticized for being difficult to test, for having weak research methods, and for overemphasizing people's ability to define social reality.

The **exchange theoretical perspective** emphasizes how people use their ____________________ to bargain in relationships to maximize their benefits ("How much can I get out of this relationship?") and to minimize their costs ("How much is this relationship going to cost me?"). According to this theory, people use their resources to bargain with others, gain advantage, and minimized being disadvantaged, even in relationships assumed to be immune from this type of "calculation." You may want to use your own intellectual skills to think of the strengths and weaknesses of this theoretical perspective.

The __________ ___________ perspective views families as a web of relationships enmeshed in the broader web of relationships that is the community or society. In this perspective, both families and societies tend to attain a somewhat stable equilibrium. Change in family system affects not only its equilibrium, but also brings about change in society's equilibrium, and vice versa. The strength of family system is to alert us to the fact that family and society are part of the web of relationships we recognize as social reality. But this is also its weakness as a theory: this generalization is so broadly and generally true that once its truth has been grasped, it is difficult to know what else can be learned from the perspective.

The **conflict/feminist perspective** emphasizes the fact that what is in one person's or group's interests are not necessarily the interests of all. This theory points out that what one person or group considers a gain may be considered as a loss by another person or group. According to this perspective, conflict is no one's "fault." Conflict just "is." Conflict theorists view conflict as a natural, normal, inevitable part of any social relationship, and as something that may have positive or negative consequences, or both. The strength of conflict theory is its ability to direct our attention to conflict as an important, inevitable part of life for couples, families, and societies. It has been criticized, however, for being too "political," too utopian, and as being more inevitable than it may be.

Social scientists have devised a number of research methods to obtain information about family relationships. Social scientists do this because although we may call upon _______________ **opinion** and experience for the beginning of an answer to the questions that interest us, everyone's personal experience is ________________. Furthermore, we have no way to be sure that our experiences are typical. Therefore, **scientific investigation** — with its various methodological techniques — is designed to provide more effective ways of gathering knowledge about the family. The purpose of **surveys, laboratory observation and experiment,** ____________________ **observation, clinicians' case studies, longitudinal studies**, and of historical and **cross-** _______________ **data** is to provide us with more accurate, reliable, and valid knowledge about the family.

Point to Ponder: At a Spring, 1996 political convention 53 resolutions were presented for consideration, debate, and vote. Many of the resolutions were passed with little or no debate. There was much agreement on many of the fiscal issues such as income tax reform, a line item veto for the President, and a balanced budget amendment.

The social issues, however, reflected more disagreement among the delegates. A resolution was presented to define the term *family* as being "a man and his wife with their offspring." Many delgates spoke in opposition to that definition. However, the motion to narrowly define the word "family" was voted on and passed.

KEY TERMS

You should be able to explain the concepts listed below. In your explanation, try to avoid using the concept you are explaining. You should be able to *give several examples* of each concept and to *explain why* each example is an example.

theoretical perspective 28
family ecology perspective 28
family policy 29
family development perspective 30
structure-functional perspective 30
social institutions 31
self-sufficient economic unit 32
monogamy 36
polygamy 36
nuclear family 36
extended family 36
experiential reality 40
agreement reality 41
interactionist perspective 35
self concept 36
identity 36
exchange theory 37
family systems theory 38
conflict perspective 39
feminist theory 39
cultural equivalent approach 42
cultural deviant approach 42
cultural variant approach 42
kin scripts framework 43
scientific investigation 41
surveys 41
experiment 44
naturalistic observation 45
case studies 46
longitudinal studies 46
heterosexism 41

COMPLETION

Complete the following sentences by selecting the correct alternative from the Key Terms listed above. Some key terms may be used more than once. Some may not be used at all. Filling in a blank may require more than one word.

1. ______________________________ refers to a type of marriage in which there is a sexually exclusive union of one man and one woman.

2. In ______________________________ , a person has more than one spouse.

3. In a narrow sense, ______________________________ is all the procedures, regulations, attitudes, and glals of government that affect families.

4. What members of a society *believe* to be true, or, ______________________________, may misrepresent the *actual* realities experienced by families.

5. In the ________________________________ theoretical perspective, society is viewed as part of families' environment, placing limitations and constraints on families, but also as opening up possibilities and opportunities for families.

6. Patterned and predictable ways of thinking and behaving—beliefs, values, attitudes, norms—that are organized around important aspects of group life and thus meet important needs or serve important social functions, are termed social ________________________________.

7. In the ____ to minority families, people in the cultural mainstream view minority families' values, beliefs, and behaviors as negative or pathological.

8. In the ________________________________ research method, samples of persons are contacted by telephone, mail, or in person, and are asked to respond to questionnaires, after which the data are analyzed and conclusions drawn.

9. The ________________________________ or theory has been criticized for overestimating the amount of shared values, cooperation and equilibrium, and overlooking or de-emphasizing the amount of conflict and stress in social institutions and structures.

10. In the research method termed ________________________________, the researcher allow us to view family behavior as it actually happens in its natural setting.

11. The ________________________________ or theory has been criticized for overestimating the degree to which family members create realities of their own making as compared to the "real" reality in which they find themselves.

12. Carefully measured, monitored, and controlled conditions are necessary if family researchers are to use the research method termed "________________________________."

13. Scientists conduct ________________________________ when they get information about individuals, families, or larger groups and do so in such a way as to make comparisons over a longer period of time.

14. A(n) ________________________________ consists of three or more generations sharing, among other things, work and resources.

KEY RESEARCH STUDIES

U.S. Bureau of the Census. The U.S. Bureau of the Census continually collects and publishes information about the U.S. population and is a major source of data on changes in Marriage and family trends. Ask your librarian to show you where and how to find these and similar reports in your library. If you do not already know where and how to use computer resources to locate this and other family information, ask the librarian.

KEY THEORETICAL PERSPECTIVES

The purpose of any theory is to help us to see the world more clearly and explain *why* things are they are, thus leading us to greater understanding. You should be familiar with the following key theories, being able to explain each briefly or at length, give examples of each theory, and answer questions about them.

ecological perspective
family development perspective
structure functional perspective
interactionist perspective
interactionist perspective
family systems theory
conflict/feminist perspective

Quotations to assess: Read the following quotations and then think about them, assessing or evaluating them in terms of the text material in this chapter. Do not worry about being right or wrong. Just read the quotation and then think about in terms of the concepts, theories, ideas, and research results covered in this chapter of the textbook.

"There never comes a point where a theory can be said to be true. The most that one can claim for any theory is that it has shared the successes of all its rivals and that it has passed at least one test which they have failed."
A. J. Ayer (1910-80), British philosopher. *Philosophy in the Twentieth Century.* ch. 4 (1982).

"Traditional scientific method has always been at the very best, 20-20 hindsight. It's good for seeing where you've been. It's good for testing the truth of what you think you know, but it can't tell you where you ought to go."
Robert M. Pirsig (b. 1928), U.S. author. *Zen and the Art of Motorcycle Maintenance*, pt. 3, ch. 24 (1974)

"Like the Pentagon, our social science often reduces all phenomena to dollars and body counts. Sexuality, family unity, kinship, masculine solidarity, maternity, motivation, nurturing, all the rituals of personal identity and development, all the bonds of community, seem 'sexist,' 'superstitious,' 'mystical,' 'inefficient,' 'discriminatory.' And, of course, they are — and they are also indispensable to a civilized society."
Robert Gilder (b. 1939), U.S. author, speechwriter, author . *Sexual Suicide*, Introduction (1973).

TRUE-FALSE

If you have studied and understand the textbook, you should be able to answer correctly the following true-false questions. Some of the questions are general and some are specific. Each is either obviously true or obviously false. None of the questions are tricky. Answer all of the questions by printing a T or F in the answer space for each question, and then check your answers with the correct answers at the end of this chapter. If you find that you miss questions, review the textbook material to discover *why* your answer is incorrect

_____ 1. Theoretical perspectives are ways of looking at reality, perspectives, or points of view.

_____ 2. The text discussed seven theoretical perspectives, two of which were the family ecology and family development perspectives.

_____ 3. The family ecology perspective explores ways in which families are interdependent with their environments.

_____ 4. The discussion of government family policy seems to be a "natural" topic for the family ecology perspective.

_____ 5. Feminist perspectives about family seem to be most consistent and to conflict very little with the ideas of structure-functional theory.

_____ 6. Viewing the family as a social institution with economic, educational, and other important tasks to perform for individual family members, for families, and for society as a whole, is most consistent with the structure-functional theory.

_____ 7. Of the world's many societies, the majority do not embrace monogamy as the preferred family type. However, romantic love provides the preferred basis for marriage even in societies where non-monogamous marriage is the norm.

_____ 8. The interaction theoretical perspective is a subcategory of the conflict theoretical perspective and is consistent with it. The interaction perspective, however, tends to focus on large-scale relationships involve relatively large numbers of persons.

_____ 9. Family system theory tends to look at the family as a whole.

_____ 10. According to the textbook, there is no "typical" family or family structure.

_____ 11. Of the three types of frameworks for studying minority families, it is the "cultural equivalent" framework that would be most consist with and in agreement with the conflict theoretical perspective of family.

_____ 12. Kin-scription is the active recruitment of family members to do kin-work.

____ 13. Surveys are by their very nature an unscientific way to study families.

_____ 14. The fact that people who present themselves for counseling may differ in important ways from those who do not is an inherent weakness in the use of clinicians' case studies.

_____ 15. Longitudinal studies obtain and analyze information about changes that happen over a considerable length of time—weeks, months, years, decades, or longer.

MULTIPLE CHOICE

If you have studied and understand the textbook, you should be able to answer correctly the following multiple choice questions. Some of the questions are general and some are specific. Print the correct alternative letter-answer in the space to the left of each question. Check your answers with the correct answers at the end of this chapter. If you understand the chapter very well, you should miss few or none of these questions. For each question you miss, review the pertinent textbook material to understand *why* your answer is incorrect.

_____ 1. Activities such as raising children reasonably, providing economic support, and giving emotional security are the main focus of _________ theoretical perspective.
  a. family systems
  b. family development
  c. structure-functional
  d. none of the above

_____ 2. The purpose all theoretical perspectives is to:
  a. move from multiple small-scale multiple explanations toward a larger, comprehensive explanation
  b. increase our understanding
  c. make research methods more reliable and valid
  d. become as abstract as possible within the limits of the topic being studied

_____ 3. How families are affected by their neighborhoods and how neighborhoods affect individual families is part of the ________ theoretical perspective.
  a. family ecology
  b. structure-functional
  c. interactionist
  d. family development

_____ 4. Which of the following is consistent with the main concerns of the family ecology theoretical perspective?
  a. the family as a child-rearing institution
  b. the conflict/feminist perspective
  c. laboratory observations and longitudinal designs
  d. development of family policy

_____ 5. Which of the following is an important part of the family development theoretical perspective?
  a. family life cycle
  b. removing blinders by means of scientific research
  c. the family as an economic unit
  d. kin-work and kin-time

_____ 6. Patty and Al find that their relationship is changing rapidly due to the arrival of their first child. Their relationships is not necessarily "worse," but it is rapidly become more complex and has changed very much. Patty and Al are in the _____ stage of the stage family life cycle.
   a. second
   b. third
   c. fourth
   d. sixth

_____ 7. The family development perspective emerged and had wide acceptance during:
   a. 1880s-1910
   b. 1920s-1920s
   c. 1930s-1950s
   d. 1960s-1980s

_____ 8. Which of the following characterized the family development perspective?
   a. It had a white, middle-class bias.
   b. It was similar to the family systems approach.
   c. It was not applicable to the study of large populations.
   d. It was not applicable to families consisting of middle-aged or older partners whose adult children had left home.

_____ 9. Joel's listened intently as his parents explained to him that the family was the most of order and predictability to social relationships, providing a sense of emotional security and a satisfying feeling of belonging. Joel, who studied his family textbook and understood it, recognized immediately that his parents would be strong supporters of the _______ theoretical perspective about families.
   a. family systems
   b. exchange
   c. interactionist
   d. functional

_____ 10. The modern family is no longer a(n):
   a. mechanism for socialization and/or social control of infants and children
   b. an important source of emotional security and support, tasks now met largely by peer groups.
   c. location for physical and psychological boundaries
   d. a self-sufficient economic unit

_____ 11. Which of the following is a FALSE idea about marriages and families in various cultures?
   a. Marriage in most cultures is supposed to be monogamous.
   b. Only a slight majority of the world's cultures allow young people to select their own marriage partners.
   c. Almost universally, what we recognize as and call "Love" is the basis for marriage.
   d. all of the above

_____ 12. Which of the following theoretical perspectives believes that our concept of self and our identity emerges within the family?
   a. developmental theory
   b. interactionist theory
   c. structure-functional theory
   d. conflict/feminist theory

_____ 13. A theoretical perspective that can show how things fit together and how "everything affects everything else" within the family is:
   a. family systems theory
   b. structure-functional theory
   c. developmental theory
   d. feminist theory

_____ 14. It is true that many adult sons are caregivers for their seriously disabled parents. It is true also that caregiving for elderly parents is enthusiastically shared by now-adult sons and daughters. However, the quotation "A son's a son till he takes a wife; a daughter's a daughter all of her life" helps draw our attention to the fact that it is daughters more than sons who are likely to be involved in direct caregiving of dependent elderly parents. Of the theoretical perspectives presented in the textbook, the one that most helps to understand this caregiving situation is:
   a. interactionist theory
   b. conflict theory
   c. family system theory
   d. family development theory

_____ 15. Ann's partner Alan is objecting to the minority family in the apartment next to theirs. He doesn't like it that the minority family doesn't take their infant to nursery school while both parents work and that they don't even have a bank account. Ann explained that the minority couple don't leave their infant alone; concerned people from their minority group church take turns taking care of the infant in its own bed/play room. Moreover, Ann explained, the minority couple is a participant in an innovative minority group economic co-op which writes checks for all depositors, including the minority couple, thus paying all their bills and meeting similar standard banking needs. Ann's explanation illustrates the ____ toward family diversity.

a. cultural deviant
b. cultural variant
c. cultural equivalent
d. cultural creativity

_____ 16. "Kin-time" refers to:

a. family norms about when certain life events or life transitions should occur
b. relatives expectations about whether or not they are being neglect by those most closely related to them
c. the fact that minority attitudes about promptness and tardiness are often different that dominant majority group attitudes
d. some minority group thinking that they have a right to spend time with family members even when it interferes with work obligations

_____ 17. A research method that frequently makes use of a "random sample" is:

a. experiment
b. survey
c. longitudinal study
d. clinician's case study

_____ 18. "Normative answers" are among the drawbacks or difficulties inhere in which of the following research methods?

a. experiment
b. survey
c. longitudinal study
d. historical data

_____ 19. Which of the following is an unavoidable factor in a longitudinal research study or project, but instead is planned and must occur if the research is longitudinal?

a. recruitment of a relatively large number of research participants
b. precise measurement, often accompanied by computer-based data analysis
c. the passage of time
d. researchers being non-judgmental about the observed behavior

_____ 20. All research tools represent:

a. the main reason for selecting a specific theory to guide the research
b. the scientific community to the general public
c. a compromise, because each research tools has pro's and con's
d. the individual research or research team's level of commitment to the research project, because some research tools require more commitment than other such tools

_____ 21. The view that divorce is the dismantling of the couple relationship that has emerged over time, with newer identities arising that are not tied to the marriage, is consistent with approach taken by _____ theories of family.

a. conflict/feminist
b. developmental
c. structure-functional
d. interactionist

_____ 22. Betty and Fred remained married as long as they were both living. They had two children who lived with them until both children finished college. They grew old together and when Betty died, Fred never remarried. This illustrates:

a. polygamy
b. monogamy
c. family linkage
d. fictive kin

_____ 23. In the United States, the belief that members of a family almost always reside together in the same household is an example of:

a. a reflexive institution
b. a healthy institution
c. a statistical generalization
d. an inaccuracy caused by "blinders"

_____ 24. Betty studied conflict by randomly assigning her student research subjects to two groups. She gave both groups the same family conflict problems to resolve, but she gave one group a mini-lecture on conflict resolution. Then she measured how long it took each group to resolve its family conflict problem. This is an example of what type of investigation?

a. survey
b. experiment
c. naturalistic observation
d. clinician's case study

_____ 25. Hector programmed the computer to select at random the names of 500 students who were then individually asked the same questions by trained interviewers. Hector then used the computer to analyze the results. Hector used which type of scientific investigation?
a. longitudinal study
b. survey
c. case study
d. experiment

Your Opinion, Please: The text presents a variety of theoretical perspectives about the family. Of these varied theoretical perspectives, which *one* do you *personally* find most explanatory? You're entitled to your opinion. What is there about that theory perspective that leads you to find it the most explanatory of those discussed in the chapter? Think your answer through, and be as specific as you can in explaining why you find the theory to be explanatory. Remember, by thinking these exercises through, you will increase both your ability to recall and to understand the theories.

SHORT-ANSWER ESSAY QUESTIONS

The following are sample short-answer essay questions — questions of the type you may be asked if your instructor uses questions like these. Even if your instructor does not use questions like these, you can help organize and consolidate your learning if you can answer these questions in a well-organized and complete manner.

Do not think that it is easier to write adequate answer for brief essays than for longer essays. Short-answer essays may be more challenging because answers are required to be both brief and complete. And after you have answered these questions below, construct some similar questions of your own devising, and then answer them. The Study Questions at the end of each textbook chapter make very good short-answer essay questions.

1. Distinguish between family development theory and interactionist theory.

2. Compare and contrast the conflict theory with exchange theory. In what ways are they similar and different to each other?

3. Explain how personal experience can act as "blinders," preventing more complete understanding of couples and families.

4. Distinguish between longitudinal research and historical data.

5. What is "kin-scription"? Give an example of it from your own experience, being sure that it is clear from your example why it is a good example.

ESSAY

The following are sample essay questions — questions of the type you may be asked if your instructor uses essay questions. Even if your instructor does not use essay questions, you can help organize and consolidate your learning if you can answer these questions in a well-organized and complete manner. Of these essay questions, the third essay question is usually the most challenging.

1. Of the various theories of family, no one theory is "the correct theory." Explain why.

2. Explore the strengths and weaknesses of each of the methods of social science discussed in the textbook.

3. Each theory discussed in the textbook lends itself better to some research methods than to others. Select *one* theoretical perspective and write an essay in which you point out (1) the research methods that seem particularly well-suited to research guided by that theory and (2) the research methods that should probably be avoided in research guided by that theory. You may find it helpful to use an example to be used in illustration as you write your essay.

ANSWERS

CHAPTER SUMMARY

structure-**functional**
conflict/**feminist** perspectives
by the **environments**
is so **general** that
how families **change**
structure-functional
interactionist perspective concentrates
people use their **resources**
the **family systems** perspective
call upon **personal** opinion and experience
participant observation
cross-cultural observation

COMPLETION

1. monogamy
2. polygamy
3. family policy
4. experiential reality
5. family ecology
6. institutions
7. cultural deviant approach
8. survey
9. structure-functional perspective
10. naturalistic observation
11. interactionist perspective
12. experiment
13. longitudinal studies
14. extended family

TRUE-FALSE

1. T
2. T
3. T
4. T
5. F
6. T
7. F
8. F
9. T
10. T
11. F
12. T
13. F
14. T
15. T

MULTIPLE CHOICE

1. C
2. B
3. A
4. D
5. A
6. A
7. C
8. A
9. D
10. D
11. D
12. A
13. A
14. B
15. C
16. A
17. B
18. B
19. C
20. C
21. D
22. B
23. D
24. B
25. B

CHAPTER 3

OUR GENDERED IDENTITIES

This chapter introduced the topic of gender and gender identity, as influenced by genetics, culture, and socialization. Increasingly, the term "sex" is used in reference to the biological facts of male and female, while the term "gender" is used to refer to how culture, society, and socialization defines identities and behavior based on sex differences. Cultural definitions of gender are discussed in terms of the interplay between the interrelationship between biology and culture in producing gender-based attitudes and behaviors. Socialization —especially in families, peer groups, and schools—play an important part in perpetuating gender identities and behaviors. The textbook explores the importance of gender identities in adult lives, examining the conflict, confusion, and hope surrounding these issues.

CHAPTER SUMMARY

Roles of men and women have changed over time, but living in our society remains a different experience for women and for men. __________ **messages** and social organization influence people's behavior, attitudes, and options. In the dominant culture of the United States, women tend to be seen as more expressive, relationship-oriented, "communal," and as having more_______________ **character traits**; men tend to be considered more agentic and as having more __________________ **character traits**. When these cultural expectations result in expected behavior patterns based on gender, ______________________________(s) are the result.

Stereotypes of African-American men and woman are more similar to each other than are those of other Americans. Generally, traditional masculine expectations require that men be confident, self-reliant, and occupationally successful and engage in "no sissy stuff." During the 1980s, the "new male" (or "liberated male") cultural message emerged, according to which men are expected to value tenderness and equal relationships with women. Traditional feminine expectations involve a woman's being a man's help-mate and a "good mother." An emergent feminine role is the successful "professional woman"; when coupled with the more traditional ones, this results in the "superwoman."

The extent to which men and women differ from one another and follow these cultural messages can be visualized as two overlapping normal distribution curves. The statistical "means" differ according to cultural expectations, but within-group variation is usually greater than between-group variation. Put in non-statistical terms, this simply means that there is more variation among males as a group —and among women as a group— that there is when males as a group are compared with females as a group. An exception is ____________________________________, evident in politics, in religion, and (although this is changing for many men and women), in the economy.

Biology interacts with culture to produce human behavior, and the two influences are difficult to separate. Sociologists give greater attention to the ______________________________ process, for which there are several theoretical explanations. Overall, however, sociologists stress the socialization process in the family, in ______________________________, and in school as encouraging gendered attitudes and behavior.

Women and men negotiate gendered expectations and make choices in a context of change at work and in relationships. Change brings mixed responses, depending on class, racial/ethnic membership, religion, or other social indicators. New cultural ideals are far from realization, and efforts to create lives balancing love and work involve conflict and struggle. But the greater equality that allows conflict between partners also creates the conditions for satisfying intimate relationships.

Point to Ponder: If female students and male students have different achievement levels in computer science courses — and there is evidence that they do — is it because male students and female students have different socialization experiences in terms of learning computer science, or is it because of inherent differences in the abilities of males and females? One study found that when female students in computer science had questions or difficulties about computers, the computer lab assistants tended to "fix" the women's computer problems *for* them rather than explain computer principles so the women could learning more and in the future possibly better solve their own computer problems. But when male students in the same courses had questions, computer lab assistants tended to *explain* computer principles to the men, thus helping the male students to learn more about computers and to be more equipped to solve their own computer problems in the future.

If male and female computer science students are found to have differing knowledge and skill levels at the end of a computer course taught this way, do you think it is because of innate or "natural" differences between males and females, or can you think of an alternative explanation?

Can you think of other areas of college/university life in which this type of learning occurs?

KEY TERMS

You should be able to explain the concepts listed below. In your explanation, try to avoid using the concept you are explaining. You should be able to *give several examples* of each concept and to *explain why* each example is an example.

gendered 52
sex 52
gender 52
gender role 52
instrumental character traits 52
expressive character traits 52
masculinities 54
femininities 54
hermaphrodites 63
assignment 63
hormones 63
interactive influence of nature and nurture 64
internalization 64
socialization 65
social learning theory 65
self identification theory 66
border work 69
androgyny 76

COMPLETION

Complete the following sentences by selecting the correct alternative from the Key Terms listed above. Some key terms may be used more than once. Some may not be used at all. Filling in a blank may require more than one word.

1. Biological and/or physiological differences between males and females are referred to as ________________________ differences.

2. The general process by which society influences people to internalize attitudes and expectations is called ______________________________________.

3. Sex _________________________________ influence sexual dimorphism: sex-related differences in body structure and size, muscle development, and voice quality.

4. Competitiveness, self-confidence, logic, and nonemotionality as allegedly "natural" for men are examples of ____.

5. In politics, religion, and in the economy, there is evidence of __________________________.

6. ____ are perceived to be more "feminine," and include sensitivity to the needs of others and the ability to express tender feelings.

7. All men are not alike. That is why it makes sense to speak of

   ______________________________.

8. As discussed in this chapter, the process of deciding that an infant will be raised as a male or raised as a female is the meaning of the term ______________________________.

9. The concepts of "good mother," "professional woman," and "superwoman" are examples of

   ______________________________.

10. The term ______________________________ refers to the social and psychological condition in which individuals can think, feel, and behave both instrumentally and expressively, showing both masculine and feminine characteristics.

11. Paying little or no attention to sexual differences as far as role behavior is concerned but instead expecting behavior to be appropriate to either sex is called______________________________.

12. Ted retorts to Monique "Just because I am male doesn't mean I am automatically and forever the head of the household but you get to do the dishes. Just because I am male doesn't mean I must be the primary wage-earner but that you have the option to decide whether or not you want to work and contributed toward paying the bills. Where is it engraved in stone that it has to be that way? I don't like it and I am not sure I can stay with you any more unless we can talk this through and come to some mutual agreement we both feel we are comfortable with." Clearly, Ted has come to the point where he is aggressively questioning a ______________________________.

KEY RESEARCH STUDIES

Money and Ehrhardt: conclusions abased on studies about hermaphrodites
Maccoby and Jacklin: innate vs. learned gender differences
Geerson's research on women's and men's role choices

KEY THEORETICAL PERSPECTIVES

Socialization and gender
Chodorow's theory of gender
Kohlberg's self-identification theory
Androgyny and egalitarianism

Quotations to assess: Read the following quotations and then think about them, assessing or evaluating them in terms of the text material in this chapter. Do not worry about being right or wrong. Just read the quotation and then think about in terms of the concepts, theories, ideas, and research results covered in this chapter of the textbook.

"Except for their genitals, I don't know what immutable differences exist between men and women. Perhaps there are some other unchangeable differences; probably there are a number of irrelevant differences. But is clear that until social expectations for men and women are equal, until we provide equal respect for both sexes, answers to this question will simply reflect our prejudices."
Naomi Weisstein (b. 1939), U.S. psychologist, author, "Woman as Nigger," in *Psychology Today* (New York, Oct. 1969).

"How beautiful maleness is, if it finds its right expression."
D.H. Lawrence, (1850-1930), British author. Sea and Sardinia, ch. 3 (1923).

"We allow our ignorance to prevail upon us and make us think we can survive alone, alone in patches, alone in groups, alone in races, even alone in genders."
Maya Angalou (b. 1928), U.S. author. Address to Centenary College of Louisiana (reported in *New York Times*, 11 March 1990)

TRUE-FALSE

If you have studied and understand the textbook, you should be able to answer correctly the following true-false questions. Some of the questions are general and some are specific. Each is either obviously true or obviously false. None of the questions are tricky. Answer all of the questions by printing a T or F in the answer space for each question, and then check your answers with the correct answers at the end of this chapter. If you find that you miss questions, review the textbook material to discover *why* your answer is incorrect.

_____ 1. Gender influence perhaps as much as three-fourths of a person's life and relationships.

_____ 2. Gender is a central organizing principle in all societies.

_____ 3. In the context of this chapter, "sex" refers only to male or female physiology, or, to the biological aspects of male/female differences.

_____ 4. "Gender" involves social roles and stereotypical cultural messages about what it means to be masculine or feminine.

_____ 5. Almost all people internalize gender-related attitudes and expectations for behavior.

_____ 6. "Instrumental character traits" are those that include such traits as logic, confidence, and competitiveness.

_____ 7. Gender differences *among* women (and *among* men) are usually greater than these differences *between* men and women.

_____ 8. "Agentic" character traits are usually expressive traits.

_____ 9. The concept of "masculinities" was constructed so that anomalies such as gay males could be include within the scope of the term; "femininities" has the same origin as pertains to lesbians.

_____ 10. The vast majority of heterosexual or "straight" men are very clear in their minds about what it means to be masculine.

_____ 11. That a person cannot be both masculine and feminine is one of the conclusions emerging from a recent study of college students' descriptions of "masculine" and "feminine."

_____ 12. On an institutional level, male dominance is especially evident in the institutional spheres of politics, the economy, and education.

_____ 13. Male chimpanzees are less dominant and aggressive than are male baboons.

_____ 14. There is less male dominance in foraging and hoe societies than in both agricultural societies and industrial societies.

_____ 15. In their study of hermaphrodites, Money and Ehrhardt found that the most crucial factor in these infants' ultimate self-perceptions was hormonal, not the presence or type of male/female genitalia.

_____ 16. Human hormonal levels are relatively unaffected by what is happening in their environment.

_____ 17. According to Chodorow's theory of gender, an infant's "primary identification" is developed with their same-sex parent, a process that is difficult to thwart or impede even when the same-sex parent cannot spend much time with the infant.

_____ 18. From nursery school through high school, teachers pay more attention to male than to female students.

_____ 19. Chicanas are more like nonHispanic white women than they are like Mexican-American men.

_____ 20. "Masculinists" are men involved in a men's movement viewing women as "the Enemy" who threaten male prerogatives and traditionally male spheres of influence.

MULTIPLE CHOICE

If you have studied and understand the textbook, you should be able to answer correctly the following multiple choice questions. Some of the questions are general and some are specific. Print the correct alternative letter-answer in the space to the left of each question. Check your answers with the correct answers at the end of this chapter. If you understand the chapter very well, you should miss few or none of these questions. For each question you miss, review the pertinent textbook material to understand *why* your answer is incorrect.

_____ 1. The textbook chapter on sex, gender roles, gender identities, and sex roles is most directly concerned with which of the following controversies?
   a. the group versus the individual
   b. the society versus the culture
   c. technology versus the individual
   d. heredity versus the social environment

_____ 2. African-American males tend to be viewed perhaps being _____ to each other in terms of expressiveness and competence than are non-Hispanic whites.
   a. equal
   b. more similar
   c. more dissimilar
   d. much more dissimilar

_____ 3. We tend to have stereotypes about all of the following *except* which one?
   a. middle-class white women
   b. gays
   c. elderly males
   d. none: we have stereotypes about all of the above

_____ 4. Which of the following is false regarding masculinities and femininities as cultural messages?
   a. Cultural messages deliver several possible masculinities and femininities as cultural messages.
   b. Cultural messages about males tend to be more specific and varied than are cultural messages about females.
   c. Cultural messages about masculinities and femininities lead to confusion and ambiguity in the socialization process.
   d. none of the above

_____ 5. Acting according to cultural expectations may be:
   a. conditional
   b. situational
   c. future-oriented
   d. dramaturgical

_____ 6. Which of these is not one of the one of the areas noted by the textbook as characterized by a high degree of male dominance?
   a. the medical institutions
   b. the religious institutions
   c. the economic institutions
   d. the political institutions

_____ 7. Approximately _____ of U.S. Roman Catholics believe that women priests in the Roman Catholic church think that women priests "would be a good thing."
   a. one-tenth
   b. one-fourth
   c. one-half
   d. two-thirds

_____ 8. Although in no age category are earnings equal between men and women, among women it is the _____ women who earn a larger proportion of what men earn.
   a. younger
   b. older
   c. widowed or divorced
   d. Hispanic

_____ 9. Which of the following is consistent with what the textbook says about those having the highest probability of disapproving of women priests?
   a. Father Anthony, who was married and divorced before entering seminary.
   b. Bishop Friedlund, who has written several books about the Church's social obligations
   c. Paul, age 8, who gets good grades in science, language, and lives with his divorced father
   d. Judie, age 28, who works in her husband's office and, unpaid, is totally responsible for keeping all business financial records

_____ 10. Money and Ehrhardt found that the most crucial factor in hermaphrodite infants' ultimate perception of themselves as boys or girls, and later as men or women, was:
   a. the sex of the physician responsible for the medical care of the infant
   b. whether the parents lived in a predominantly rural or predominantly urban area
   c. educational level of parents
   d. assignment

______ 11. One study of the effects of testosterone on male behavior found that males with more testosterone may be less likely to ____ and more like to ______.

a. watch television; actively participate in sports
b. enjoy solitary sports; participate in team activities
c. marry; divorce
d. enjoy leisure; become "workaholics"

______ 12. Theresa is now ten years old. Since birth, she has always lived with her single-parent father and has no female relatives living in the region of the nation where she and her father live. Nevertheless, Theresa thinks of herself as a girl, and has both values, behaviors, and attitudes that probably most "average people" would consider traditionally feminine. Which of the following best explains why/how this happened?

a. Money and Ehrhardt's theory
b. Chodorow's theory of gender
c. Kohlberg's self-identification theory
d. Martin's impression vulnerability theory

______ 13. Teddy is great at 8th grade computer science. Most of the boys in class have tend to do better in the class than do the girls. A possible explanation for this phenomenon may be that boys tend to have _____ that tend to be different that those of girls.

a. toys
b. lower left-quadrant brains
c. time-use patterns
d. language skills

______ 14. Sexual harassment and rape on campus may be especially strongly associated:

a. membership in fraternities and/or sororities
b. having lower than average grades
c. living off-campus rather than in dormitory settings
d. female students' presents in campus environments dominated by males (physics, engineering, physical education)

______ 15. A study of white, African-American, Latina and Asian-American girls found that Latina and Asian-American girls were very like to:

a. de-emphasize education in preference for domestic roles
b. be very quiet in the classroom and less likely to be noticed by teachers
c. get the highest grades in the class and criticized for it by other students
d. be perceived by teachers as a vague but definite threat to the smooth running of the classroom environment

_____ 16. In a 1989 Yankelovich telephone poll of 1,000 women across the country, it was found that _______ of the women polled said the movement "looks down on women who do not have jobs."
   a. 6 percent
   b. 13 percent
   c. 22 percent
   d. 35 percent

_____ 17. Chicanas who support insurgent feminism are in favor of:
   a. ultra-traditional roles for Mexican-American females
   b. revolutionary or radical restructuring of society to end all forms of oppression
   c. planned patterning of feminine expectations to more closely mirror those of white females
   d. focusing on the educational and occupational systems as ways of improving role options available to Chicanas

_____ 18. Which of these is not one of the types of men's movements noted by Kimmel?
   a. masculinists
   b. male survivalists
   c. antifeminists
   d. profeminists

_____ 19. "Androgyny" is a term with Greek origins, meaning literally:
   a. "male" and "female"
   b. hermaphroditic
   c. a masculinized female
   d. a feminized male

_____ 20. As used in the textbook, a man who is androgynous:
   a. expresses positive qualities traditionally associated with both males and females
   b. has difficulty fitting in with male peer groups or with female peer groups, due mainly to a kind of psychological "averaging" of their personality and the behavior that flows from it
   c. can express male attitudes and behaviors, but only with difficulty, and can rarely sustain these attitudes and behaviors over a sustained period of time
   d. seems to be exceptionally attractive to women but has difficulty sustaining male friendships

_____ 21. According to the textbook, partly responsible for people having mixed feelings and conflicts about gender roles is which of the following?
   a. transitional mixes of hormones and other biochemical blood-substances
   b. educational institutions' persistence in using out-of-date textbooks depicting gender expectations
   c. our increased options for defining and expressing our gender identities
   d. the increase of single parenthood and children raised in dual-income families

_____ 22. Which of the following *is* in the forefront of attention of persons who are observers of movements related to changing definitions of gender roles?
  a. consciousness-raising movements among teenagers
  b. gay men and lesbians who are organizing to achieve social goals related to their sexual orientation
  c. many men's ambivalence, confusion, and anger about changing gender expectations
  d. elected state and federal legislators who are racing to make our laws and governmental regulations consistent with the "social agenda"

Your Opinion, Please: Some people think that it works out best for all concerned if "men are men" and "women are women," with each sex displaying gender roles that most people in society expect. But some people think that such notions are too narrow and keep people from having new experiences, from exploring new aspects of personality that remained unexamined. What are your views on this issue? Are you in support of a world in which gender roles are high predictable or would you prefer a world in which people are to a high degree androgynous?

If you support androgyny at all, are there some spheres of daily life where you feel it is more appropriate than others? Any areas in which you feel it is less appropriate? Why?

SHORT-ANSWER ESSAY QUESTIONS

The following are sample short-answer essay questions — questions of the type you may be asked if your instructor uses questions like these. Even if your instructor does not use questions like these, you can help organize and consolidate your learning if you can answer these questions in a well-organized and complete manner.

Do not think that it is easier to write adequate answer for brief essays than for longer essays. Short-answer essays may be more challenging because answers are required to be both brief and complete. And after you have answered these questions below, construct some similar questions of your own devising, and then answer them. The Study Questions at the end of each textbook chapter make very good short-answer essay questions.

1. To what extent do individual men and women fit gender stereotypes and expectations?

2. Summarize what is known about made dominance in agricultural societies.

3. If males dominate in the political sphere—and they do—in what ways and by what mechanisms do they dominate?

4. Summarize the ways in which play and games are related to socialization to gender expectations, identities, and roles.

5. Summarize what the differences Orenstein found between African-America, Latina, white, and Asian-American girls in middle school.

6. According to the text, what has been the response of black women about committing themselves to feminist goals?

7. Describe the goals, interests, and difficulties of Chicana cultural nationalist feminists.

8. Distinguish between "masculinists" and "anti-feminists."

ESSAY

The following are sample essay questions — questions of the type you may be asked if your instructor uses essay questions. Even if your instructor does not use essay questions, you can help organize and consolidate your learning if you can answer these questions in a well-organized and complete manner. Of these essay questions, the third essay question is usually the most challenging.

1. What are the origins of gender differences and inequalities according to *society*-based explanations? Be specific and support your answer with such facts as help to enhance the persuasiveness of your essay.

2. In an organized, well-written essay, explore how culture and biology interact to produce, intensify, and perpetuate gender differences. Support your essay with specific terms, theories, and research results wherever possible and appropriate.

3. Write an essay in which you explore how aggressiveness illustrates the interaction between culture and biology to produce gender differences. Support your essay with specific terms, theories, and research results wherever possible and appropriate.

4. Distinguish between social learning theory and self-identification theory of acquiring gender attitudes, expectations, and behaviors. In your essay be sure to point out the similarities and differences of these two explanations.

5. In schools, what are the major sources of students being socialized into learning gender expectations, and how does each source contribute to the overall result that gender expectations are among the things students learn at school?

6. According to the text, schools help perpetuate and intensify gender attitudes and expectations. Write an essay in which you explore how the factors mentioned in textbook seem evident at the college or university in which you are enrolled. A fruitful way to approach this essay might be to write first a mechanism reported in the text, and then follow that with something equivalent you have noted at college/university. Be specific, using terms appropriately and citing research results where appropriate to support the ideas in your essay.

7. Select one of the following minority groups and write an essay in which you summarize the chapter's information about that minority group and gender realities: African-Americans women, or Hispanic women (Latinas, Chicanas).

ANSWERS

CHAPTER SUMMARY

gendered messages and social organization
more **expressive** character traits; men tend
instrumental character traits; when these
gender roles are the result
male dominance is the result
socialization process
in the family, in **play and games**, and in school

COMPLETION

1. sex
2. socialization
3. hormones
4. instrumental character traits
5. male dominance
6. expressive character traits
7. masculinities
8. assignment
9. femininities
10. androgyny
11. androgyny
12. gender role

TRUE-FALSE

1. F
2. T
3. T
4. T
5. T
6. T
7. T
8. F
9. F
10. F
11. F
12. F
13. T
14. T
15. F
16. F
17. F
18. T
19. F
20. F

MULTIPLE CHOICE

1. D
2. B
3. D
4. B
5. B
6. A
7. D
8. A
9. B
10. D
11. C
12. C
13. A
14. A
15. B
16. D
17. B
18. B
19. A
20. A
21. C
22. C

CHAPTER 4

LOVING OURSELVES AND OTHERS

This chapter examines love as an aspect of personal, cultural, and social life. It begins by exploring what it is about society today that makes love so important to so many people. The chapter then investigates what love is and what love isn't. Love is seen as a phenomenon with its own requirements in terms of the characteristics of the individuals involved. Because love is a process that involves disclosure and attendant risk, the textbook suggests that healthy self-esteem is a prerequisite to loving. After explaining that love is not a fact but a process, the chapter closes with some comments about keeping love.

CHAPTER SUMMARY

In an impersonal society, _____________ provides an important source of fulfillment and intimacy. Genuine loving in our society is rare and difficult to learn. Our culture emphasizes self-reliance and rationality as central values, as exemplified in our expectation that people leave their friends and relatives to move to job-determined new locations because it is an economically efficient way of organizing production and management. In the face of all this impersonality, aridity, and paucity of intimate human contact, most people search for at least one caring person with whom to share their private time. Loving and being loved have important consequences for emotional and physical well-being.

In the **triangular theory of love**, Sternberg uses the three dimensions of **intimacy**, _______________, and **commitment**, to generate a typology of love, one of which types, called_________________________**love,** involves all three components. John Lee lists six love styles: **eros**, or passionate love; **storge**, or companionate, familiar love; **pragma**, or pragmatic love; _________________ , or altruistic love; **ludus** or love play; and __________________________, or possessive love.

Despite its importance, however, love is often misunderstood. It should not be confused with **martyring** or ______________________________. There are many contemporary love styles that indicate the range of dimensions love-like relationships—not necessarily love—can take.

Self-esteem seems to be a prerequisite for loving. In relationships where the is emotional **interdependence**, love probably involves acceptance of oneself and others, a sense of empathy, and a willingness to let down barriers set up for self-preservation. **A-frame**, **H-frame**, and _____ **-frame** relationships are some of the forms love can take. People discover love; they don't simply find it or have it strike like a thunderbolt. Put different, love is not an event; love is an unfolding process. Reiss's _______________________ **of love** sets forth four basic stages of the love process: **rapport**, ______________________________ , **mutual dependency**, and **personal need fulfillment**.

Since partners need to keep on sharing their thoughts, feelings, troubles, and joys with each other, love is a continual process. Discovering love implies a process, and to develop and maintain a loving relationship requires mutual ________________, requiring time and trust.

> **Point to Ponder:** Perhaps most people in our society think that romantic love is the appropriate basis for marriage. And yet we know that not societies base marriage on individual choice. Some societies have "arranged marriage." In such societies, parents, siblings, or hired match-makers arrange to pair people up with each other. For the couple, love is largely left until later in marriage, if at all. What are the positive and negative consequences of arranged marriage, from your personal point of view. So you think that the positive consequences outweigh the negative ones? Try to imagine yourself in an arranged marriage. Do you like what you see? Why? Or, why not?

KEY TERMS

love, 84
emotion, 84
legitimate needs, 85
illegitimate needs, 85
intimacy, 87
commitment, 87
sexual intimacy, 89
psychic intimacy, 89
Sternberg's triangular theory of love, 90
intimacy (Sternberg's theory of), 90
passion (Sternberg's theory), 90
commitment (Sternberg's theory), 90
consummate love, 90
love styles, 91
Eros, 91
storge, 91
pragma, 91
agape, 91
ludus, 91
mania, 91
martyring, 92
manipulating, 93
symbiotic relationships, 93
self-esteem, 94
narcissism, 94
dependence, 96
independence, 96
interdependence, 97
A-frame relationships, 97
H-frame relationships, 97
M-frame relationships, 97
wheel of love, 98
self-disclosure, 99

COMPLETION

Complete the following sentences by selecting the correct alternative from the Key Terms listed above. Some key terms may be used more than once. Some may not be used at all. Filling in a blank may require more than one word.

1. A(n) ______________________ is a strong feeling, arising without conscious mental or rational effort, that motivates an individual to behave in a certain way.

2. ______________________ are those arising in the present rather than out of deficits or failures accumulated in the past.

3. ______________________ spring from feelings of self-doubt, unworthiness, and inadequacy.

4. According to Sternberg's triangular theory of love, when intimacy, passion, and commitment are all involved, the result is the type of love called______________________.

5. ______________________ refer to the various distinctive "personalities" that love-like relationships can take.

6. The type of love that is characterized by intense emotional attachment and powerful sexual feelings or desires is termed ______________________.

7. The term ______________________ refers to an affectionate, companionate style of loving.

8. ______________________ is the kind of love involving an emphasis on the practical side of human relationships, emphasizing economic and emotional security.

9. Often called altruistic love, ______________________ emphasizes unselfish concern for the beloved.

10. The type of love that emphasizes playfulness and the humorous or amusing aspects of the love relationship is what is meant by ______________________.

11. An insatiable need for attention and affection alternating between euphoria and depression is characteristic of a ______________________love style.

12. ______________________________ involves maintaining relationships by giving others more than one receives in return, usually with good intentions, but seldom with the feeling that genuine affection is being received.

13. ______________________________ means seeking to control the feelings, attitudes, and behavior of the partner by subtle, indirect ways rather than by straightforwardly stating one's case or position about the matter at hand.

14. Martyrs and manipulators are often attracted to each other, forming what counselor John Crosby calls a(n) ______________________________.

15. ______________________________ refers to an evaluation a person makes and maintains of her- or himself that expresses attitudes of approval or disapproval, success or failure, worth or unworthiness, and similar ideas.

16. ______________________________ is essentially another word for selfishness.

17. The concept of ______________________________ refers to a general reliance on another person or on other people for continual support or assurance, along with subordination.

18. The first stage in the wheel theory of the development of love is ______________________________.

19. ______________________________ refers to self-reliance, self-sufficiency, sometimes including separation or isolation from others.

20. When we label a relationship ______________________________, we imply that the people involved have high self-esteem and make strong commitments to each other.

21. In ______________________________ relationships, the partners stand self-sufficient and virtually alone, with neither influenced much by the other.

22. In ______________________________, each partner has high self-esteem, experiences loving as deep emotion, and is involved in mutual influence and emotional support.

23. The stages of rapport, self-disclosure, mutual dependence, and personality need fulfillment characterize the ______________________________.

24. In Reiss's wheel theory of the development of love, ______________________________ follows self-revelation and precedes personality need fulfillment.

25. In ___________________________________ the partners have strong couple identity but little individual self-esteem.

26. Clarissa needed and could seem to function or find a reason for living without Pedro. Pedro relied upon and couldn't imagine going through life without Clarissa. Each fulfilled and seemed to make life possible for the other. Pedro became terminally ill and before he died of his illness, he and Clarissa followed through on their earlier agreement. They committed mutual suicide. Of the A-, H-, and M-frame relationships, theirs seems to have been a(n)_____________________.

KEY RESEARCH STUDIES

Cancian: Do men and women "care" differently?

KEY THEORETICAL PERSPECTIVES

Sternberg: the triangular theory of love
Crosby: three types of interdependence relationships
Lee: six love styles
Reiss: the wheel of love

Quotations to assess: Read the following quotations and then think about them, assessing or evaluating their meanings or evaluating them in terms of the text material in this chapter. Do not worry about being right or wrong. Just read the quotations and then think about them in terms of the concepts, theories, ideas and research results covered in this chapter of the textbook.

"I've only been in love with a beer bottle and a mirror."
Sid Vicious (1957-79), British punk rocker. ***Sounds*** (London, 9 Oct. 1976).

"If love ... means that one person absorbs the other, then no real relationship exists any more. Love evaporates; there is nothing left to love. The integrity of self is gone."
Ann Oakley (b. 1944), British sociologist, author. ***Taking It Like a Woman***, "Love: Irresolution" (1984)".

"Love is the direct opposite of hate. By definition it's something you can't feel for more than a few minutes at a time, so what's all this bullshit about loving somebody for the rest of your life?"
Judith Rossner (b. 1935), U.S. author, ***Nine Months in the Life of an Old Maid***, pt. 2 (1969).

"We're born alone, we live alone, we die alone. Only through our love and friendship can we create the illusion for the moment that we're not alone."
Orson Welles (1915-1984), U.S. filmmaker, actor, producer. Definition of loneliness contributed by Welles to Henry Jaglom, ***Someone to Love*** (1985).

"Today I begin to understand what love must be, if it exists.... When we are parted, we each feel the lack of the other half of ourselves. We are incomplete like a book in two volumes of which the first has been lost. That is what I imagine love to be: incompleteness in absence."
Goncour, Edmond de (1822-96) and Jules de (1830-70), French writers. ***The Goncourt Journals*** (1888-96; repr. in ***Pages from the Goncourt Journal***, ed by Robert Baldick, 1962) entry for 15 Nov. 1859.

TRUE-FALSE

If you have studied and understand the textbook, you should be able to answer correctly the following true-false questions. Print a T or F in the answer space for each question, and then check your answers with the correct answers at the end of this chapter.

_____ 1. Modern society in the United States is often characterized as impersonal.

_____ 2. Ideally, lovers support and encourage each other's personal growth.

_____ 3. Men tend to see sex as one of several ways of communication on an emotional level.

_____ 4. Cancian maintains that men are as loving as women are but are not recognized as such because men tend to express love in nonverbal ways, and the recognized language of love in the U.S. is primarily verbal.

_____ 5. "Legitimate needs" are those that enjoy the approval of recognized social institutions, such as family, friends, and work associates.

_____ 6. "Illegitimate needs" are those needed by lovers who were themselves born out of wedlock or to unmarried mothers.

_____ 7. Intimacy, passion, and commitment are all parts of consummate love in the triangular theory of love.

_____ 8. The pragma type of love emphasizes the practical element in human relationships.

_____ 9. Agape refers to a type of love typified by unselfish concern for the beloved.

_____ 10. Denise can't stop thinking about Freddie, even for an hour. She has almost nothing else on her mind except trying to satisfy her intense emotional and sexual attraction to Freddie, who maddeningly remains just beyond her reach. This an example of the love style called "mania."

_____ 11. The text feels that martyring is one of the highest types of love relationships.

_____ 12. Manipulating is an activity that is part of a normal, natural, and productive part of the love relationship, regardless of the general public's negative view of manipulation. Manipulation is common, normal, usual, and healthy.

_____ 13. "Manipulating" is not a type of love, but "martyring is a type of love.

_____ 14. Being in love with another person involves changing yourself to be more like what the other person wants you to become.

_____ 15. A-frame relationships are characteristic of couples in which the individuals have a strong couple identity but low levels of self-esteem.

_____ 16. In H-frame relationships, the partners stand virtually alone and isolated, with little or no couple identity and little emotionality.

_____ 17. In Reiss's wheel of love, the first stage is "rapport."

_____ 18. The first moment Frank met Annette, he shared with her some things about himself, things most people weren't aware of. But even so, he and Annette seemed to find this first meeting comfortable and pleasant. Frank and Annette had reversed the first two stages of Reiss's wheel of love.

_____ 19. Mutual dependence is the fourth stage in Reiss's wheel of love.

_____ 20. To develop and maintain a loving relationship requires self-disclosure.

MULTIPLE CHOICE

If you have studied and understand the textbook, you should be able to answer correctly the following multiple choice questions. Some of the questions are general and some are specific. Print the correct alternative letter-answer in the space to the left of each question. Check your answers with the correct answers at the end of this chapter. If you understand the chapter very well, you should miss few or none of these questions. For each question you miss, review the pertinent textbook material to understand *why* your answer is incorrect.

_____ 1. One of the reasons love may be so important in the United States is that much of every life is:
a. perplexing
b. impersonal
c. ambiguous
d. unrewarding

_____ 2. Some who have studied modern society are warning against the _____ in our social world. The text seems to point to this issue as one of the reasons love is so important to people today.
a. extreme rationality and individuality
b. low levels of goal orientation and persistence
c. low levels of logical planning and problem solving
d. inordinately high levels of religious influence

_____ 3. The concept of illegitimate needs directly involves which of the following?
a. failures from the past that remain "unfinished business"
b. requests by the loved one that just can't be met by any reasonable person
c. needs that have the potential to snowball into something more
d. needs that most people would view as somewhat deviant

_____ 4. According to the text, men in our society have great difficulty in:
a. bridging the gap between liking and loving
b. falling in love with women of their own social class
c. establishing relationships between their lovers and their own parents
d. communicating about love verbally.

_____ 5. Cancian argues that in our society _____ are made to feel primarily responsible for the endurance or success of love.
a. mothers
b. psychological and psychiatric counselors and therapists
c. women
d. "activity" industries, such as cruise boats, restaurants, and ski lodges

_____ 6. The case study about Sharon and Gary, who discover love after twenty-five years, is about a couple who:
a. marry, separate, and then rediscover love
b. discover love after twenty-five years of singlehood
c. are—separately—widowed, but then discover that they are still capable of loving
d. marry, divorce, and then each finds love with a different partner

_____ 7. Which of these is a friendly, affectionate, and companionate style of love?
a. storge
b. eros
c. pragma
d. agape

_____ 8. The playful aspect of love is what is meant by which of the following styles of love?
a. eros
b. mania
c. pragma
d. ludus

_____ 9. Mike and LaPrelle feel strong love for each other. Perhaps the most notable characteristic of their love relationships is that they have such a good time, get such a "kick" out of one another, and spontaneously laugh and joke as they go about their activities. These are hallmarks of a(n) __________ love relationship.
a. storge
b. eros
c. agape
d. ludus

_____ 10. Unselfish concern for the beloved is the basic idea in which of the following styles of love?
a. eros
b. pragma
c. ludus
d. agape

_____ 11. Moodiness, jealousy, and insatiable need are signs of which of the following love styles?
a. ludus
b. pragma
c. mania
d. agape

_____ 12. It is characteristic of martyrs that they think that:
a. love is something worth dying for
b. it is better to be wanted as a victim than not to be wanted at all
c. there is nobility in making the beloved do what you want him/her to do
d. one should not compromise in order to sustain a love relationship

_____ 13. "Manipulating" is best illustrated by which of the following?
a. "I understand what you feel. Now let me tell you how I feel." (Said softly.)
b. "I see what you have done. This is what I'm gonna do now." (Openly stated.)
c. "If you will do what I want you do, only then will I do what you want me to do." (hinted, but not said openly)
d. "If you do not do as I wish, then I will do as I threatened." (openly stated)

_____ 14. Manipulators, like martyrs, do *not* believe that:
a. love really exists; they are cynical about love
b. loving is a normal, desirable phenomenon; they see it as desirable
c. the partner is worth loving; they seriously undervalue the partner
d. they are lovable or that others can really love them

_____ 15. Family counselor John Crosby has discussed _______ in which each partner depends on the other for a sense of self-worth.
a. noblesse oblige
b. benign parasitism
c. mutual dependence neurosis
d. symbiotic relationships

_____ 16. The feelings people have about their own worth are called their feelings of:
a. self-concept
b. self-esteem
c. personal value
d. personal assessment

_____ 17. The text states that _____ is a prerequisite for loving.
a. self-esteem
b. narcissism
c. androgyny
d. spontaneous erotic skill or sexual expertise

_____ 18. The two members of a couple confirm that they are involved in a couple relationship. They feel it is satisfying. Basically, however, each member is self-sufficient and is in some ways isolated and independent from the other person in the relationship. This situation describes a(n) _______ relationship.
a. A-frame
b. H-frame
c. K-frame
d. M-frame

_____ 19. A thorough reading of the text's discussion of "discovering" love is to see love viewed as which of the following:
a. a process or sequence
b. a rare but marvelous coincidence
c. an objective fact
d. an encounter with a unique kind of love

_____ 20. People _____ love; they don't just find it.
a. invent
b. prolong
c. discover
d. defend themselves against or thwart

_____ 21. The second stage in Reiss's wheel theory of love is:
a. mutual dependence
b. personality need fulfillment
c. rapport
d. self-revelation

_____ 22. Peeling away outer layers of the self, outer layers of "protective coloration" we use to shield our vulnerability, and in the process showing more and more of the "real" or authentic self, is part of the _____ stage in Reiss's wheel theory of love.
a. first
b. second
c. third
d. fourth

_____ 23. Bruce and Jane have some things they share in common but they also have some individual activities. The things they share are very important to them: they divide the activities of these tasks, enjoy these tasks most when the other person is there with them to take part in it, and really would prefer to not do the activity rather than do it without the other person being there to take part. This is part of the _____ stage in Reiss's wheel theory of love.
a. first
b. second
c. third
d. fourth

_____ 24. Pat and Carolyn have come to feel that they need to be together always because each feels that the other person completes them and makes them a better person because of the relationship. This part of the ______ stage in Reiss's wheel theory of love.

a. first
b. second
c. third
d. fourth

_____ 25. Lisette and Claude didn't feel comfortable with one another from the moment they were introduced. They just couldn't get through those first moments comfortably. According to Reiss's wheel theory of love, they were unable to establish ______ when they met.

a. self-revelation
b. mutual dependency
c. rapport
d. personality need fulfillment

Your Opinion, Please: How do you know when you are "in love" with someone? A few questions that come to mind are these: Can you be "in love" with a person and yet dislike him or her as a person? Is there more than one person in the world with whom you could be "in love"? Is it possible to be "in love" with more than one person at a time? After you have given your own opinion, think about what the authors of the text would answer to each of the above questions, based on what they have written in this chapter. If you are feeling bold, ask these same questions of someone with whom you have a relationship. Next. do you feel comfortable sharing with that person your own answers to the questions? Why or why not?

SHORT-ANSWER ESSAY QUESTIONS

The following are sample short-answer essay questions — questions of the type you may be asked if your instructor uses questions like these. Even if your instructor does not use questions like these, you can help organize and consolidate your learning if you can answer these questions in a well-organized and complete manner.

Do not think that it is easier to write adequate answer for brief essays than for longer essays. Short-answer essays may be more challenging because answers are required to be both brief and complete. And after you have answered these questions below, construct some similar questions of your own devising, and then answer them. The Study Questions at the end of each textbook chapter make very good short-answer essay questions.

1. Distinguish between legitimate needs and illegitimate needs that love might satisfy.

2. What are the main parts of the triangular theory of love? How does consummate love fit into this theory Explain carefully, briefly, but completely.

3. Sometimes lovers say they would be willing to "do anything" for the sake of their beloved. Is this the same thing as "martyring"? Why or why not?

4. In Reiss's wheel theory of love, why does the rapport stage precede the self-revelation stage? Explain in terms of process.

5. Define *both* M-frame and A-frame relationship. After defining both, give an example of either *one* of these two types of relationships, making sure it is clear from your example *why* it is a good example.

ESSAY

The following are sample essay questions — questions of the type you may be asked if your instructor uses essay questions. Even if your instructor does not use essay questions, you can help organize and consolidate your learning if you can answer these questions in a well-organized and complete manner. Of these essay questions, the third essay question is usually the most challenging.

1. Explain what love is and what it is not.

2. In what way(s) is self-esteem a prerequisite to love or loving? That is, does self-esteem seem to be required before one can fully love?

3. Compare and contrast Sternberg's triangular theory of love with Reiss's wheel theory of love. Do they have similar goals? Do they address similar or different issues? Explain in a well-organized essay.

ANSWERS

CHAPTER SUMMARY

love provides an important source of fulfillment
intimacy, **passion**, and commitment
one of which, **consummate** love
agape, or altruistic love
mania, or possessive love
martyring or **manipulating**
H-frame, or **M-**frame
wheel of love
rapport, **self-disclosure**
mutual **self-disclosure**

COMPLETION

1 emotion
2. legitimate needs
3. illegitimate needs
4 consummate love
5. love styles
6. eros
7. storge
8. pragma
9. agape
10. ludus
11. mania
12. martyring
13. manipulating
14. symbiotic relationship
15. self-esteem
16. narcissism
17. dependence
18. rapport
19. independence
20. interdependent
21. H-frame
22. M-frame
23. wheel of love
24. mutual dependency
25. A-frame
26. A-frame

TRUE-FALSE

1. T
2. T
3. F
4. T
5. F
6. F
7. T
8. T
9. T
10. T
11. F
12. F
13. F
14. F
15. T
16. T
17. T
18. T
19. F
20. T

MULTIPLE CHOICE

1. B
2. A
3. B
4. D
5. C
6. A
7. A
8. D
9. D
10. D
11. C
12. B
13. C
14. D
15. D
16. B
17. A
18. B
19. A
20. C
21. D
22. B
23. C
24. D
25. C

CHAPTER 5

OUR SEXUAL SELVES

This chapter examines the development of sexual orientation and sexual expression as affected by historical circumstances of culture and society, examines sexuality throughout marriage, including sex as a pleasure bond in relationships. The chapter concludes with an examination of the ways in which sexual expression and HIV/AIDS affect each other.

CHAPTER SUMMARY

Social attitudes and values plan an important role in the formation of sexual expression people find comfortable and enjoyable. This point applies to many aspects of sexuality, including even our sexual orientation — whether we prefer a partner of the same or opposite sex. Despite decades of conjecture and research, it is still unclear whether it is genetic or socially conditioned. The 1980's and 1990s have witnessed political and other challenges by gay activists(and a few others) to heterosexism and one of its consequences, ________________________.

Whatever one's sexual orientation, sexual expression is negotiated with cultural messages about what is sexually permissible, even desirable. In the United States, the cultural messages have moved from one that encouraged patriarchal sex, based on ________________________________ and reproduction as its principal purpose, to a message that encourages sexual expressiveness in myriad ways for both genders equally.

African Americans may be more sexually expressive and less inhibited than other Americans. Four standards of nonmarital sex are ________________________, permissiveness with affection, permissiveness without affection, and the ________________________________ —diminished somewhat since the 1960s, but still alive and well.

Marital sex changes throughout life. Young spouses tend to place greater emphasis on sex than do older mates. But, while the frequency of sexual intercourse declines over time and the length of a marriage, some 27 percent of married persons over age 74 are sexually active and having sex about four times a month. Making sex this kind of ______________________ bond, whether married or not, involves cooperating in a nurturing, caring relationship. To fully cooperate sexually, partners need to develop high ________________________, to break free from restrictive gendered stereotypes, and to communicate openly.

The text also discussed ____________________ disease, with some focus on how the disease affects relationships, marriages, and families. That general discussion led to the consideration of how the social institutions of religion and politics today have merged somewhat to legislate for more restrictive values regarding sex. The New ______________________________ organization or movement advocates "family" values that include celibacy outside marriage, traditional roles for women, and heterosexism. Whether or not one agrees with the major agenda of that organization or movement, there are certain guidelines for personal sexual responsibility that we should all heed.

Points to Ponder: If you made a list of the ten things you personally think have the greatest effect on personal relationships these days, would HIV/AIDS be on your list? It would be on many people's list. Think of the difference it has made in sexual expression or behavior.

That difference would be greater if more people changed their behavior in ways consistent with protecting themselves and their partner(s) in light of the HIV/AIDS problem. However, many have *not* changed their behavior in these ways .

According to the text, "In one study 35 percent of men reported lying to a partner in order to have sex with her, while 60 percent of women thought they had been lied to" (p. 138). That was a study of what *men* said. But just as a matter of curiosity, what percent of *women* do you think lie about *their* sexual history to a partner they are about have sex with? What percent of men do you predict would report they thought *they* had been lied to in this way?

Some have said that one can do without taking precautions for safer sex only in cases where both partners are free of the disease and are *100 percent certain* that not only they but their partner is monogamous or, sexually faithful. Questions: Can one *ever* be 100 percent certain that one's partner has been 100% sexually faithful? And what are the implications of your answer to that question for your own sexual behavior? (Take a few moments to think about this. It's an important issue.)

KEY TERMS

You should be able to explain the concepts listed below. In your explanation, try to avoid using the concept you are explaining. You should be able to *give several examples* of each concept and to *explain why* each example is an example.

sexual orientation 107
heterosexuals 107
homosexuals 107
interactionist perspective
on human sexuality 108
evolutionary psychology 108
survival of the fittest 108
patriarchal sexuality 109
expressive sexuality 109
homophobia 111
heterosexism 111
abstinence 117
permissiveness with affection 117
permissiveness without affection 117
double standard 118
representative samples 120
habituation 123
pleasure bond 124
sexual responsibility 125
pleasuring 125
spectatoring 125
holistic view of sex 126
HIV/AIDS 128
safer sex 131
New Christian Right 136

COMPLETION

Complete the following sentences by selecting the correct alternative from the Key Terms listed above. Some key terms may be used more than once. Some may not be used at all. Filling in a blank may require more than one word.

1. The sexual standard of ____________________________ maintains that regardless of the circumstances, nonmarital intercourse is wrong for both women and men.

2. ____________________________________ refers to whether an individual prefers a partner of the same or opposite sex.

3. Darwin proposed that all species evolve according to the principle of the ____.

4. The practice of ____ faces the fact that condoms can be broken .

5. ______________________ is a viral disease that destroys the immune system.

6. Having high self-esteem allows us to engage in ___________________________, or, spontaneously doing what feels good at the moment and letting orgasm happen (or not), rather than working to produce it.

7. According to the ___________________________________, women's sexual behavior must be more conservative than men's, and in its original form it meant that women shouldn't have sex before or outside of marriage, whereas men could.

8. The view that women and men negotiate and are influenced by the "sexual scripts" they learn from society is called the____.

9. Decreased interest in sex that results from the increased accessibility of a sexual partner and the predictability in sexual behavior with that partner over time, is called______________________.

10. The practice of emotionally removing oneself from a sexual encounter only to watch, analyze, and judge one's sexual ability, performance or productivity is called ___________________________.

11. Taking a(n) _________________________________ refers to looking at sex as something that is an extension of (or taking place within the context of) the entire relationship between the partners, thus seeing as more than simply a physical relationship.

12. The sexual standard of ____ permits nonmarital intercourse for both men and women equally, provided they have a fairly stable, affectionate relationship.

13. Doris's approach to sexual activity includes relaxing and trying to make her partner "feel good." And by feel good she means enjoying sensations, not feeling any need to hurry or to slow down, enjoying feelings and emotions for their own sake, and not worrying about whether or not orgasm occurs. Doris's approach to sexual activity as described obviously includes that which the textbook terms ____.

14. Roberto can't seem to relax and enjoy sexual or erotic activity. No matter how hard he tries, he keeps having the feeling that somebody is watching him. He now realizes that the person "watching" him has been himself, judging his sexual "performance," comparing to what he thinks others might expect it to be, evaluating it in terms of what he has internalized as being the "right kind of way" for a man to behave sexually and to make his partner respond in a sexual sense. Roberto has been engaging in ____________________________________.

KEY RESEARCH STUDIES

Research results of 1993 New York Times poll regarding how the public views gay issues.
Kinsey, Pomeroy, and Martin research on sexual behavior.
National Opinion Research Center (NORC) 1992 survey on sexual behavior.
Masters, Johnson, and Kolodny research about sex as a pleasure bond.
National Centers for Disease Control research about HIV/AIDS.

KEY THEORIES

Interactionist perspective on human sexuality

Quotations to assess: Read the following quotations and then think about them, assessing or evaluating their meanings or evaluating them in terms of the text material in this chapter. Do not worry about being right or wrong. Just read the quotations and then think about them in terms of the concepts, theories, ideas and research results covered in this chapter of the textbook.

"If I were asked for a one line answer to the question "What makes a woman good in bed?" I would say, 'A man who is good in bed.'
Bob Guccione (B. 1930), U.S. publisher. Interview in Wendy Leigh, *Speaking Frankly* (1978).

"Sex is a conversation carried out by other means. If you get on well out of bed, half the problems of bed are solved."
Peter Ustinov (b. 1921), British actor, writer, director. Interview in Wendy Leigh, *Speaking Frankly* (1978).

"All men are homosexual, some turn straight. It must be very odd to be a straight man because your sexuality is hopelessly defensive. It's like an ideal of racial purity."
Derek Jarman (b. 1942), British filmmaker, artist, author. *At Your Own Risk: A Saint's Testament*, "1940's" (1992).

"The modern erotic ideal: man and woman in loving sexual embrace experiencing simultaneous orgasm through genital intercourse. This is a psychiatric-sexual myth useful for fostering feelings of sexual inadequacy and personal inferiority. It is also a rich source of 'psychiatric patients.' "
Thomas Szasz (b. 1920), U.S. psychiatrist. *The Second Sin*, "Sex" (1973)

"We do not go to bed in single pairs; even if we choose not to refer to them, we still drag there with us the cultural impedimenta of our social class, our parents' lives, our bank balances, our sexual and emotional expectations, our whole biographies—all the bits and pieces of our unique existences."
Angela Carter (1940-1992), British author. *The Sadeian Woman*, "Polemical Preface" (1979).

"... strung out and spotty, you wriggle and sigh
and kiss all the fellows and make them all die."
Les Murray (b. 1938), Australian poet. *Little Boy Blue.*

"Sex is the mysticism of materialism and the only possible religion in a materialistic society."
Malcomb Muggeridge (1903-1990), British broadcaster. Television broadcast, BBC1, 21 Oct. 1965. Quoted in: *Muggeridge Through the Microphone*, "The American Way of Sex" (1967).

TRUE-FALSE

If you have studied and understand the textbook, you should be able to answer correctly the following true-false questions. Print a T or F in the answer space for each question, and then check your answers with the correct answers at the end of this chapter.

_____ 1. Many physiological parts of the male and female genital systems are either alike or directly analogous.

_____ 2. Heterosexual and homosexual are both types of "sexual orientation."

_____ 3. Ted is a male who is sexually attracted to other males, and he behaves accordingly. Ted is a homosexual.

_____ 4. Orgasm is important for women primarily from a psychological point of view; for men, orgasm's importance stems primarily from a physiological need.

_____ 5. "Expressive sexuality" is an need for both sexes, but research seems to indicate that it is a greater need in females than in males.

_____ 6. There is no such thing as a purely vaginal orgasm.

_____ 7. More married women are experiencing orgasm, and they are doing so more often.

_____ 8. In the NORC poll of 1,154 Americans (1993), 44 percent saw being gay as a choice.

_____ 9. Gay male and lesbian sexuality tends to be more person-centered; heterosexual sexuality tends to be more body-centered.

_____ 10. More equal gender expectations facilitate better sex.

_____ 11. Abstinence, or celibacy, receives support from counselors and feminists as a positive choice, as something that ought to remain a valid and respected option.

_____ 12. Andre is very liberal and has high self-esteem. He doesn't feel threatened by letting Ann experiment sexually with other men. He continues to feel love and warmth toward her and feels little or no jealousy. Andre illustrates what the text terms "permissiveness with affection."

_____ 13. The double standard refers to the expectation by at least one of the partners that at least two acts of sexual intercourse or the equivalent should take place during any sexual encounter.

_____ 14. The double standard is basically a dead issue for young people and, basically, does not exist as a standard because a slight majority actually overachieve in this regard.

_____ 15. According to the emergent new double standard, women are expected to be more sexually active and men less so.

_____ 16. The authors of the textbook conclude that Americans average having sex about once a week.

_____ 17. According to research by Call, Sprecher, and Schwartz, 27 percent of respondents over age 74 reported having had sex within the previous month.

_____ 18. Angela fantasizes often about having sex with Ray. She can't seem to get him out of her mind or her bed. Angela tries to manipulate Ray into having sex with her as often as she can manage it. Angela illustrates what the textbook means by the term "habituation."

_____ 19. Angela and Ray find their enjoyment of sex is increased when they watch another couple having sexual intercourse, either "live" on videotape. They are also "turned on" sexually by being watched by another couple or by videotaping their own sexual behavior. This illustrates what the textbook means by the term "spectatoring."

_____ 20. The textbook reports that the use of condoms is a way to make certain that one does not become exposed to HIV during an act of intercourse.

MULTIPLE CHOICE

If you have studied and understand the textbook, you should be able to answer correctly the following multiple choice questions. Some of the questions are general and some are specific. Print the correct alternative letter-answer in the space to the left of each question. Check your answers with the correct answers at the end of this chapter. If you understand the chapter very well, you should miss few or none of these questions. For each question you miss, review the pertinent textbook material to understand *why* your answer is incorrect.

_____ 1. The patterns of sexual response are:
   a. very similar in men and women
   b. basically the same for all age groups
   c. basically the same for almost all racial groups—but due to a few undeniable hereditary biological differences, not for all racial groups
   d. enhanced by the use of mild depressants (pot, weed, marijuana) but ironically reduced by the use of mild stimulants (crack cocaine, speed, heroin)

_____ 2. Sexual orientation may be a(n):
   a. choice, as about 74 percent of Americans believe
   b. inherited fact, and not a choice, as about 74 percent of American believe
   c. continuum rather than a dichotomy
   d. none of the above

_____ 3. Regarding human sexuality, the textbook takes primarily a(n) _________ perspective.
   a. structure functionalist
   b. conflict/feminist
   c. interactionist
   d. systems

_____ 4. The clitoris is involved in what proportion of female orgasms?
   a. only about 5 percent; most are vaginal in origin or cause
   b. as many as 50 percent; the other half are largely vaginal in origin or cause
   c. between 60-80 percent; the remainder are largely vaginal in origin or cause
   d. virtually all

_____ 5. A "bi-phobe" is a person who has a dislike, aversion to, and often a dread of:
   a. both heterosexual and homosexuals
   b. bisexuals
   c. his/her own sexual preference and that of the partner, regardless of what those preferences are
   d. none of the above

_____ 6. About ____ percent of Americans in the 1993 Gallup poll thought that gay male and lesbian relation are morally wrong.
   a. 35
   b. 55
   c. 70
   d. 78

_____ 7. In the process of constructing and taking on gay male and/or lesbian identities, "sensitization" occurs:
   a. before puberty
   b. at the time of first gay or lesbian sexual experience
   c. when sexual orientation first leads to trouble or "problems"
   d. when one's significant others become aware of the gay or lesbian sexual orientation

_____ 8. In research about four kinds of sexual relations, lesbian relationships were found to be the:
   a. least likely to lead to trouble or "problems"
   b. likeliest to have occurred at the earliest ages
   c. likeliest to have occurred at later ages
   d. least sexualized

_____ 9. Lesbian theorist Marilyn Frye writes that __________ is a topic of considerable ambiguity and/or confusion?
a. whether lesbian sexual enjoyment ought to be modeled after male sexual enjoyment
b. whether lesbian sexuality ought to be "politicized"
c. what it means to say that lesbians "are sexual beings"
d. what it means to say that lesbians "have sex"

_____ 10 About what percent of Americans give religious or moral beliefs as their reason(s) for opposing sex outside marriage?
a. 7 percent
b. 24 percent
c. 63 percent
d. 83 percent

_____ 11. According to the double standard, women's sexual behavior must be:
a. more conservative than men's
b. equal in frequency but allowably less enjoyed than men's
c. more certain to be enjoyed than that of men, and this responsibility falls primarily on the male partner
d. more certain to be enjoyed than that of men, but this responsibility falls equally on men and women

_____ 12. Some research cited in the text found that ______ are less _______ about sex.
a. African Americans; Puritanical
b. African Americans; secretive
c. Asian-Americans; open and more embarrassed
d. Mexican-Americans; frequent but more "involved"

_____ 13. Some research cited in the text found that _____ had more premarital and extramarital sex partners than _______ had.
a. middle-aged women; middle-aged men
b. middle-aged gay men; young lesbians
c. African American men; white men
d. white-collar females; blue-collar females

_____ 14. The textbook states that sex surveys must take must take greater pains to ____ than do researchers in general, because many persons consider the topic sensitive or taboo, or because some people maybe be either too willing or too reluctant to participate in the research.
a. make measurable hypotheses
b. gather data that can be handled using computer-based data analysis
c. strive for representative samples
d. attempt to use the experiment or laboratory observation

_____ 15. About ___ percent of spouses under age 25 report having had sex at least once during the previous month.

a. 99.4
b. 95
c. 88.2
d. 61

_____ 16. According to the textbook, age affects frequency of sexual intercourse; also important is:

a. ethnicity
b. geographical region in which the household of the respondent is located
c. presence/absence of children in the couple's family history
d. level of health

_____ 17. Sexual frequency declines:

a. sharply after about the first year of marriage
b. sharply during age 50-60, but increases between age 60-70
c. gradually after the birth of children, and remains lower throughout the duration of the marriage
d. when conflict is initiated by female spouses, but not when conflict is initiated by male spouses

_____ 18. __________ gives us the freedom to search for new pleasures.

a. permissiveness with affection
b. permissiveness without affection
c. narcissism
d. high self-esteem

_____ 19. About what percent of all HIV/AIDS cases in the U.S. are the result of receiving infected blood from transfusions (although since 1985 donated blood is rigorously screened for HIV/AIDS).

a. 2 percent
b. 7 percent
c. 13 percent
d. 21 percent

_____ 20. When Masters and Johnson's book *The Human Sexual Response* was published, the negative letters they received surpassed the more favorable ones by a ratio of about:

a. 2 to 1
b. 4 to 1
c. 6 to 1
d. 9 to 1

Your Opinion, Please: Someone once said that when sex is good, then it is 5 percent of a marriage. But when it is bad, then it is 95 percent of a marriage. What do you think such a viewpoint means? If you think it is true, why do you think it might be true? Try to give specific reasons for your viewpoint(s).

SHORT-ANSWER ESSAY QUESTIONS

The following are sample short-answer essay questions — questions of the type you may be asked if your instructor uses questions like these. Even if your instructor does not use questions like these, you can help organize and consolidate your learning if you can answer these questions in a well-organized and complete manner.

Do not think that it is easier to write adequate answer for brief essays than for longer essays. Short-answer essays may be more challenging because answers are required to be both brief and complete. And after you have answered these questions below, construct some similar questions of your own devising, and then answer them. The Study Questions at the end of each textbook chapter make very good short-answer essay questions.

1. Briefly but completely distinguish between homosexual, bi-sexual, and heterosexual.

2. Characterize or describe the relationship between increased age and sexual behavior, as explained in the textbook. You may find that making a list of research findings is an effective way to do this.

3. Makc a list of ways people can help protect themselves against getting AIDS/HIV.

ESSAY

The following are sample essay questions — questions of the type you may be asked if your instructor uses essay questions. Even if your instructor does not use essay questions, you can help organize and consolidate your learning if you can answer these questions in a well-organized and complete manner. Of these essay questions, the third essay question is usually the most challenging.

1. In what ways is sexual response related to self-esteem, pleasure, communication, and sharing?

2. Describe sexuality through the various stages of marriage. Be sure to use specific concepts and research results where appropriate to give additional credibility to your answer.

3. Is there a need for society—or its agents—to structure or in other ways shape the human sexual response? If yes, why? If no, why not? Explain fully.

4. Discuss ways in which politics and sex education are related. Be complete and be specific as well as general in your answer. (Make valid generalizations, but be sure to back them up with details and/or facts.)

5. Discuss the various ways in which HIV/AIDS has affected families. Be complete and be specific as well as general in your answer. (Make valid generalizations, but be sure to back them up with details and/or facts.)

ANSWERS

CHAPTER SUMMARY

consequences, **homophobia**
based on **male dominance**
discussed **HIV/AIDS** disease
and the **double standard**

kind of **pleasure** bond
develop high **self-esteem**
sex are **abstinence**
The New **Christian Right**

COMPLETION

1. abstinence
2. sexual preference
3. survival of the fittest
4. safer sex
5. HIV/AIDS
6. pleasuring
7. double standard
8. interactionist perspective on human sexuality
9. habituation
10. spectatoring
11. holistic view of sex
12. permissiveness with affection
13. pleasuring
14. spectatoring

TRUE-FALSE

1. T
2. T
3. T
4. F
5. F
6. T
7. T
8. T
9. F
10. T
11. T
12. F
13. F
14. F
15. F
16. T
17. T
18. F
19. F
20. F

MULTIPLE CHOICE

1. A
2. C
3. C
4. D
5. B
6. B
7. A
8. D
9. D
10. D
11. A
12. A
13. C
14. C
15. B
16. D
17. A
18. D
19. A
20. D

CHAPTER 6

BEING SINGLE: ALONE AND WITH OTHERS

This chapter examines what social scientists know about singles, looking at reasons more people are single today and discusses changing cultural attitudes about singlehood. The chapter also explores the variety of singles and of single lifestyles, ending by point out that the distinction between marriage and singlehood has become blurred in recent years.

CHAPTER SUMMARY

Since the 1960s the number of singles has risen. Much of this increase is due simply to rising numbers of young adults, who are typically single. Although there is a growing tendency for these young adults to postpone marriage until they are older, this is not a new trend but rather a return to a pattern that was typical early in the twentieth century.

One reason people are postponing marriage today is that increased job and lifestyle opportunities may make marriage less attractive. Also, the low ____________________ has caused some women to postpone marriage or put it off entirely. And attitudes toward marriage and singlehood are changing so that now being single is viewed not so much as deviant but as a legitimate choice.

Singles can be classified according to whether they freely choose this option for the indefinite future (________________________________), or would prefer to marry someday but for the present have decided to remain single. (________________________________). Singles can also be classified according to whether they would prefer to be married as soon as possible, but the opportunity hasn't presented itself yet (________________________________), or whether they want to marry, but they find themselves unmarried and for whatever reason are likely to remain that way (________________________________).

Singles used to live in cities often because employment and leisure opportunities drew them there, but recently more singles have moved to the suburbs. More and more singles are living in their parents' homes; this is usually at least partly a result of economic constraints. Some singles have chosen to live in communal or group homes. A substantial number of unmarrieds (about 5 percent) are cohabiting or living together. Some are heterosexual couples, and some are gay and lesbian couples; this chapter compares the interpersonal patterns of these relationships. Finally, some women share a man in a ________________________ arrangement.

Although research consistently finds marrieds physically and psychologically healthier and happier than singles, this situation is slowly changing as singlehood allows the sexual expression once reserved for marriage and as marriage fails to guarantee the security that it once did. The limits that marriage puts on individuality seem more constraining to today's Americans.

However one chooses to live the single life, maintaining supportive social networks is important.

Point to Ponder: If someone important to you is sick or dies, do you think that you ought to be able to take time off from work to attend to personal concerns? This time off might or might not be paid. In either case, most people have no problem arranging for some time off when their husbands, wives, sons, or daughters are sick or in need of help. But many singles sometimes find it difficult to get employer approval for time off from work to help out a roommate or person sharing his/her apartment when the need arises. The problem, apparently, is their marital status of singlehood. Policy in many instances does not match personal realities in our society.

Should policies be changed so that singlehood does not result in discrimination—often discrimination with serious economic consequences? Why, specifically, do you answer this question as you do?

To get you started, consider the following possible situation:

> A family member's permission is needed to withdraw life support from a patient who is now in a "persistent vegetative stage" coma, but instead of having any family around, the only person anything like "family" is the other woman with whom the comatose patient has been in acommitted lesbian relationship for 15 years. Should her lover be permitted to sign the consent form, given the absence of any "family" member?

Try to think of similar situations in which the same problem or dilemma arising from policies that don't take singlehood into account, or that specific restrict benefits to married persons and/or their children.

KEY TERMS

You should be able to explain the concepts listed below. In your explanation, try to avoid using the concept you are explaining. You should be able to *give several examples* of each concept and to *explain why* each example is an example.

sex ratio 107
sex ratio 147
voluntary temporary singles 155
voluntary stable singles 155
involuntary temporary singles 155
involuntary stable singles 156
communes 160
cohabitation 161
domestic partner 167
common law marriage 167
double standard of aging 167
continuum of social attachment 172

COMPLETION

Complete the following sentences by selecting the correct alternative from the Key Terms listed above. Some key terms may be used more than once. Some may not be used at all. Filling in a blank may require more than one word.

1. Divorced, widowed, and never-married are subcategories of the term ______________________.

2. Some cities and corporations have recognize the status of __________________________ for partners who have co-residence, economic cooperation, commitment and a sense of personal loyalty.

3. Betty is age 43, has never married, has never wanted to marry, and wants to remain single through her life, and is thus an example of those termed ____________________________.

4. Singles engaging in ___________________________________, or living together, has gained widespread acceptance over the past thirty years.

5. Nanette, Felicia, Susan, Ann, Broderick, Bob, Ben, and Jack share the available space in a large house, their personal items, their money, cooperate in household tasks, share costs of household purchases, and seem to be committed to making their innovative arrangement "work" for them as long as they can manage to stay together. When one of person leaves it is hoped that a suitable person can be found to take their place in "the arrangement." The arrangement seems to be a(n) __________________________.

6. Jane has more and stronger emotional ties to her high school friend Ed than she has for her biological kin, and Ed feels emotional ties to Jane that are as strong as those he has to his own family. One of the reasons that Jane, Ed, *and their families* are ambivalent and confused about their relationships may be that our culture has rigid categories of relationships when what may be needed is some awareness of a(n) ____.

7. The ____________________________ is the number of men per 100 women in a group, organization, society, or other collectivity.

8. The status of ______________________________________ is a legal doctrine under which persons who live together for a certain period of time, and/or hold themselves out to the community as husband and wife, and/or who say to others using words of the present tense that they are husband and wife, are in fact "married"—as married as if they had gotten married "the regular way."

9. Kenneth likes marriage and he loved his recently deceased wife. He dislikes being single and hopes some day soon to find someone else he loves, with whom he would like to live, and to marry. Kenneth is the type of single termed a(n)

   ____.

10. Both Karen and Abraham are age 72, recently widowed, relatively affluent, well-education, and have health and energy and want to get back into the mainstream of social life. Abraham already has three women friends, has been invited to meetings by several civic groups, and has been approached about his willingness to serve as a business consultant, while Karen would have welcomed such involvements but seems to have difficulty obtaining them. A concept that is probably operating here is the ____________________________________.

KEY RESEARCH STUDIES

You should be familiar with the main question being investigated and the research findings or research for the following studies:

U.S. Census and similar studies regarding singles in the U.S. population
Staples: research on singlehood among African Americans
Characteristics of types of singles and their domestic arrangements

KEY THEORIES

Staples: typology on types of singles
Ross: continuum of social attachment

Quotations to assess: Read the following quotations and then think about them, assessing or evaluating their meanings or evaluating them in terms of the text material in this chapter. Do not worry about being right or wrong. Just read the quotations and then think about them in terms of the concepts, theories, ideas and research results covered in this chapter of the textbook.

"Certainly the best works, and of greatest merit for the public, have proceeded from the unmarried, or childless men."
Francies Bacon (1561-1626), English philosopher, essayist, statesman. *Essays*, "Of Marriage and Single Life."

"He travels fastest who travels alone, and that goes double for she. Real feminism is spinsterhood."
Florence King (b. 1936), U.S. author. *Reflections In a Jaundiced Eye*, "Spinsterhood Is Powerful" (1989).

"Nowadays, all the married men live like bachelors, and all the bachelors like married men."
Oscar Wilde (1854-1900), Anglo-Irish playwright, author. Lady Hunstanton, in *A Woman of No Importance*, act 2.

"There is simply no way for a woman to live alone. Oh, she can get along financially perhaps (though not nearly as well as a man), but emotionally she is never left in peace. Her friends, her family, her fellow workers never let her forget that her husbandlessness, her childlessness—her selfishness, in short—is a reproach to the American way of life."
Erica Jong (b. 1942), U.S. author. Isadora Wing, the narrator in *Fear of Flying*. ch. 1 (1979).

"The most threatened group in human societies as in animal societies is the unmated male: the unmated male is more likely to wind up in prison or in an asylum or dead than his mated counterpart. He is less likely to be promoted at work and he is considered a poor credit risk."
Germaine Green (b. 1939), Australian feminist writer. Sex and Destiny, ch. 2 (1984).

TRUE-FALSE

If you have studied and understand the textbook, you should be able to answer correctly the following true-false questions. Print a T or F in the answer space for each question, and then check your answers with the correct answers at the end of this chapter.

_____ 1. In the past thirty years, singlehood has been on the increase in the United States.

_____ 2. Singles have increased in absolute numbers partly because the population as a whole has grown., and there are simply more people in the age bracket less likely to be married.

_____ 3. There are two demographic categories of singles: never-married, and divorced.

_____ 4. Children born during the Depression and who came to maturity in the 1950s seem to a large extent to have relatively low commitment to traditional family values.

_____ 5. The sex ratio is a measure of the ratio of actual sexual encounters compared to the opportunity for sexual encounters.

_____ 6. The proportion of older people in the population has increased, a fact that is directly related to the status of widowhood.

_____ 7. Married people tend to place a higher value on marriage, children, and love, whereas singles value friends and personal growth more.

_____ 8. The proportion of young people leaving their parental home to establish a nonmarital family (cohabitation or single parenting) has risen since about 1930.

_____ 9. Sociologist Frances Goldscheider has called women's growing lack of interest in marriage the real revolution of the last twenty years.

_____ 10. Experts are divided on the likely futures of singlehood and marriage.

_____ 11. College-educated non-Hispanic white women in the southeastern states have difficulty finding potential mates of similar educational background.

_____ 12. College-educated black women's singlehood reminds us that single status may be in part the result of a structural lack of options.

_____ 13. The concept of a "voluntary stable single" refers to a mentally secure individual who choose singlehood because they can cope very well on their own and have no insecurities that propel them toward marriage.

_____ 14. Susan's husband died suddenly. Enjoying marriage and being only age 22, Susan hopes to be married again relatively soon. She is an involuntary temporary single.

_____ 15. Single men have a more difficult time than do married men finding employment in the United States.

_____ 16. Discrimination on the basis of marital status is not part of federal civil rights law.

_____ 17. The majority of young unmarried people live in apartments.

_____ 18. Sexual arrangements in communes range from monogamy to sexual sharing.

_____ 19. People who are common law married ordinarily are or have been cohabiting.

_____ 20. Cohabitors are likelier than married partners to say they are less satisfied with their relationships.

_____ 21. By law, homosexual marriage is not legal in the United States.

_____ 22. Gays and lesbians are very likely to adopt traditional masculine and feminine roles in their relationships.

_____ 23. According to the boxed insert "The Legal Side of Living Together — Some Advice," there are no legal restrictions against an unmarried couple's opening a joint bank account.

_____ 24. The distinction between married and single is no longer very clear.

_____ 25. The "compensation phenomenon" refers to a way to compensate or "make up" for the problems due to the double standard of aging.

_____ 26. In contrast to work-class black men's friendships, upwardly mobile black men's friendships tend to be warmer and more intimate.

_____ 27. Maintaining what the textbook calls supportive social networks is important for singles is not important for marrieds.

MULTIPLE CHOICE

If you have studied and understand the textbook, you should be able to answer correctly the following multiple choice questions. Some of the questions are general and some are specific. Print the correct alternative letter-answer in the space to the left of each question. Check your answers with the correct answers at the end of this chapter. If you understand the chapter very well, you should miss few or none of these questions. For each question you miss, review the pertinent textbook material to understand *why* your answer is incorrect.

_____ 1. The U.S. Bureau of the Census recognizes how many types or categories of singles?
- a. two
- b. three
- c. four
- d. five

_____ 2. African-Americans are _______ than non-Hispanic whites to be divorced.
- a. neither more nor less likely
- b. slightly more likely
- c. much more likely
- d. slightly less likely

_____ 3. In which of the following sex ratios are there more males than females?
- a. + 75
- b. 76
- c. 98
- d. 102

_____ 4. The U.S. divorce rate is now:
- a. continuing to increase
- b. currently unstable at a high level
- c. showing a slow decrease over the last 6 years
- d. currently stable

_____ 5. The tendency in the U.S. today is to view being single as:
- a. an option
- b. social deviance
- c. regrettable and hopefully temporary
- d. a hardship on the individual and on her/his kin

_____ 6. The _________ enables men to escape pressure to respond to changes women are making in their lives and in women's expectations of men.
- a. unequal gender balance in the judiciary and legal system
- b. patriarchal law enforcement system
- c. psychological readiness of women to "cave in"
- d. uneven sex ratio

_____ 7. A researcher reported that black men more than non-Hispanic white men worried that after they married they would:
   a. have difficulty meeting the economic demands of marriage relationships
   b. be able to maintain a monogamous relationship
   c. not be able to hang out with their friends
   d. find their newly formed family displacing their kin as a primary reference group

_____ 8. Some of the nontraditional partner choices made by well-educated black women are in large part explainable by:
   a. relatively fewer eligible black men whose characteristics are suitable for them
   b. relatively lower pressure to marry or cohabit with anyone
   c. relatively higher pressure to marry or cohabit with anyone
   d. none of the above

_____ 9. David likes being married. He was saddened and disorganized when his wife died. However, at age 32, he wants to be married again when the right woman comes along. This situation illustrates which of these?
   a. voluntary temporary single
   b. voluntary stable single
   c. involuntary stable single
   d. none of these

_____ 10. Which of these illustrates a temporary voluntary single?
   a. Patrick, who is single and intends to remain that way until he is "trapped" by some woman's pregnancy and her attendant threat surround that situation
   b. Carolyn, a professional woman who wants to marry but is waiting for "the right man" to come along before she marries
   c. Jack, who wants a wife but, given his many negative characteristics, has accepted the fact that he will probably never marry again
   d. Margaret, whose husband deserted her because she wouldn't change her ways to suit him

_____ 11. Regarding job opportunities, unmarried women may be _________ than men because of their single status.
   a. likelier to have to move geographically to get a job
   b. more disadvantaged
   c. much more disadvantaged
   d. less disadvantaged

_____ 12. The percentage of young unmarried adults living at home with a parent or parents has:
   a. increased slightly since 1960
   b. decreased slightly since 1990
   c. increased moderately since 1960
   d. increased dramatically since 1990

_____ 13. Parents tend to "get in the way of" or "interfere" with __________ who live at home.
a. single women
b. young single men, who have difficulty tolerating this interference
c. single men who are in their 20s, who tolerate this interference smilingly
d. single men who are middle aged, but who in fact welcome the interference as a sign of "caring" and as a way to have more "goods and services" at home

_____ 14. Some significant number of persons from each of the following categories *except* which *one* have found that "living together" or cohabiting without being legally married can be economically advantageous?
a. persons age 16-25
b. persons age 26-40
c. persons over age 65
d. none of the above

_____ 15. Cohabitors are more likelier than others to:
a. live in the northeastern part of the country and least likely to live in the northern midwest
b. be their parents only child
c. vote either *not at all* or in *every* election
d. be nontraditional in many other ways

_____ 16. Lesbian couples are likelier than gay males to have met through:
a. friendship networks
b. co-worker or co-professional (on-the-job) relationships
c. educational or vocational training settings
d. informal interaction at gay bars

_____ 17. It is illegal for unmarrieds to:
a. file a joint income-tax return
b. reciprocally name one another as beneficiaries of life insurance polices
c. live together but have the signature of only one of them on the lease
d. open joint charge accounts

_____ 18. An alternative household form found to suit the needs and preferences of some Black Americans is one in which:
a. two or more women, each in her own household, has relationships with several men
b. two or more men share the same woman who lives in her own household
c. two or men share the same women, moving from the household of one woman to another
d. two or more women, each in her own household, share the same man

_____ 19. Which of the following categories is likeliest to be left without a partner?
  a. older women
  b. Roman Catholic and/or Baptist women
  c. middle-aged and older gay men
  d. the well-educated

_____ 20. The newly divorced:
  a. tend to throw themselves into a frantic round of social activities
  b. tend to have a honeymoon or "pink cloud" period
  c. tend to suffer from depression
  d. have a statistically significant tendency toward sexual promiscuity

_____ 21. Perhaps the greatest challenge to unmarried individuals of both sexes is:
  a. meeting their economic needs at a sufficient level
  b. finding safer sex in a world where sexual expression is becoming a risky proposition
  c. finding a sense of personal security and safety when their perception is, realistically, that they are more at risk than married persons in this regard
  d. developing strong social networks with people who can provide them with positive, nurturing relationships

Your Opinion, Please: Some governments give tax and other incentives to married people to encourage the formation of married households. Why? Because in some countries the birth rate has fallen below the replacement level so that the population is actually shrinking at a too-rapid rate and tilting the age structure (suddenly, more older than younger people). Also, since family members often care for one another in times of poor health and illness, a responsibility and expense that may increasingly have to be borne by government in these nations.

Do you think that the perceived needs of government justify the kind of policy that gives economic and other advantages to married people to encourage married households with children? Specifically, why or why not?

SHORT-ANSWER ESSAY QUESTIONS

The following are sample short-answer essay questions — questions of the type you may be asked if your instructor uses questions like these. Even if your instructor does not use questions like these, you can help organize and consolidate your learning if you can answer these questions in a well-organized and complete manner.

Do not think that it is easier to write adequate answer for brief essays than for longer essays. Short-answer essays may be more challenging because answers are required to be both brief and complete. And after you have answered these questions below, construct some similar questions of your own devising, and then answer them. The Study Questions at the end of each textbook chapter make very good short-answer essay questions.

1. Define and give an example of the following: "commune" and "cohabitation"

2. Briefly but completely explain how the sex ratio affects opportunities and decisions regarding singlehood?

3. Summarize the research results or research findings about the friendships of African-American men.

4. Why do many young singles in the United States choose to continue to live with their parents? Be as specific as possible, within the space permitted for your answer here.

5. Make a list of the special financial problems facing persons who are legally unrelated but who live together.

6. Summarize research results or findings about the income and residential patterns of singles.

ESSAY

The following are sample essay questions — questions of the type you may be asked if your instructor uses essay questions. Even if your instructor does not use essay questions, you can help organize and consolidate your learning if you can answer these questions in a well-organized and complete manner. Of these essay questions, the third essay question is usually the most challenging.

1. In what ways is sexual response related to self-esteem, pleasure, communication, and sharing?

2. Describe sexuality through the various stages of marriage. Be sure to use specific concepts and research results where appropriate to give additional credibility to your answer.

3. Is there a need for society—or its agents—to structure or in other ways shape the human sexual response? If yes, why? If no, why not? Explain fully.

4. Discuss ways in which politics and sex education are related. Be complete and be specific as well as general in your answer. (Make valid generalizations, but be sure to back them up with details and/or facts.)

5. Discuss the various ways in which HIV/AIDS has affected families. Be complete and be specific as well as general in your answer. (Make valid generalizations, but be sure to back them up with details and/or facts.)

ANSWERS

CHAPTER SUMMARY

the low **sex ratio**
this option (**voluntary stable singles**)
single nevertheless (**voluntary temporary singles**)
remain single (**involuntary temporary singles**)
marry someday (**involuntary stable singles**)
polygamous arrangement

COMPLETION

1. single
2. domestic partners
3. voluntary stable single
4. cohabitation
5. commune
6. continuum of marital attachment
7. sex ratio
8. common law marriage
9. involuntary temporary single
10. double standard of aging

TRUE-FALSE

1. T
2. T
3. F
4. F
5. F
6. T
7. T
8. T
9. T
10. T
11. F
12. T
13. F
14. T
15. T
16. T
17. T
18. F
19. T
20. T
21. T
22. F
23. T
24. T
25. F
26. F
27. F

MULTIPLE CHOICE

1. B
2. B
3. D
4. B
5. A
6. D
7. C
8. A
9. D
10. B
11. D
12. D
13. A
14. D
15. D
16. A
17. A
18. D
19. A
20. C
21. D

CHAPTER 7

CHOOSING EACH OTHER

This chapter explores the social factors that affect choosing partners, examines marital stability, and at some discernible patterns by which individuals in our society develop commitments to each other. After reviewing variables or factors that influence the bargaining power persons have as the enter the marriage market, the authors examine the factors that work together to narrow the pool of eligibles. Courtship' several facets are explored, including patterns of increased acquaintance, courtship, cohabitation and marriage. The chapter considers the factors that affect the stability of early and late marriage, and ends by listing and reviewing the reasons for which persons choose to marry.

CHAPTER SUMMARY

Our distinctly American connection of love and marriage is unique to our modern culture. Historically, marriages were often arranged in the ______________________________, as business deals. Many elements of the basic exchange (a man's providing financial support in exchange for the women's childbearing and child-rearing capabilities, domestic services, and sexual availability) remain.

What attracts people to each other? Two important factors are homogamy and ______________________________. Some elements of homogamy are ______________________, social pressure, feeling at home with each other, and the fair exchange. Three patterns of courtship familiar in our society are dating, getting together, and ______________________________.

Besides homogamy and the degree of intimacy developed during courtship, two other factors related to the success of a marriage are a couple's age at marriage and their reasons for marrying. People who marry ________________________________ are less likely to stay married, and there are several negative reasons for marrying that can lead to unhappiness or divorce.

If potential marriage unhappiness can be anticipated, breaking up before marriage is by far the best course of action., however difficult it seems at the time. A certain number of courting relationships will end this fashion.

But many people will go on to marry, perhaps for one or more of the twelve reasons during the chapter's closing pages.

Point to Ponder: In the United States, some states allow a person to marry a first cousin, while other states prohibit such marriage. Similarly, some states do not allow adopted brothers and sisters to marry. What do states accomplish by these restrictions? What, specifically, are your reasons for your point of view, whatever it is?

KEY TERMS

You should be able to explain the concepts listed below. In your explanation, try to avoid using the concept you are explaining. You should be able to *give several examples* of each concept and to *explain why* each example is an example.

courtly love 183
marriage market 183
dowry 183
imaging, 198
exchange theory 183
marriage gradient 187
pool of eligibles 188
homogamy 188
endogamy 188
exogamy 188
heterogamy 188
interracial marriage 193
hypergamy 193
hypergamy 193
theory of complementary need 196
stimulus-values-roles (SVR) theory of courtship, 198
sex ratio 186
dating, 199
date rape, 200
getting together 202
two-stage marriage 204
cohabitation 204
"Linus blanket" 206
emancipation 206

COMPLETION

Complete the following sentences by selecting the correct alternative from the Key Terms listed above. Some key terms may be used more than once. Some may not be used at all. Filling in a blank may require more than one word.

1. Relationships based on ______________________________ required a great deal of idealization and were not necessarily sexually consummated.

2. A(n) ______________________________ is a sum of money or property brought to the marriage by the female..

3. Viewing marriage in terms of bargaining, a shopping place, and resources, is a central part of what is called the ____.

4. The ____ is the number of men to women in a society or in a subgroup of society.

5. The idea that people who date or are going together come to the relationship with their assets and liabilities and make the best deal they can is part of what is meant by the __________________ theory or perspective.

6. Those persons who have the personal, social, and other characteristics to make them an appropriate marriage choice can be termed a(n)____.

7. ____ exists when people marry others of similar race, age, education, religious background, and social class.

8. ____ occurs when people marry within their own social group.

9. ____ refers to marriage to one who belongs to a higher socioeconomic class than one's own.

10. ____ occurs when people marry outside their social group.

11. ____ occured when Kaitlin, a fifty-four year old, high school drop-out, white, Irish Catholic, married Freddie, an twenty-eight year old, college graduate, Hispanic, Baptist .

12. ____ can be a courtship process in which groups of women and men congregate at a part or share an activity, but there is relatively little pressure to relate to any one member of the opposite sex.

13. When unrelated persons of the opposite sex live together without being married, the relationship is termed ____________________________________.

KEY RESEARCH STUDIES

You should be familiar with the main question being investigated and the research findings or research for the following studies:

Norval Glenn: interreligious marriages
Research based on data from U.S. Census, U.S. National Center for Health Statistics, National Survey of Families and Households, and similar sources.

KEY THEORIES

exchange theory
the marriage gradient
theory of complementary needs
S-V-R—stimulus-values-roles

Quotations to assess: Read the following quotations and then think about them, assessing or evaluating their meanings or evaluating them in terms of the text material in this chapter. Do not worry about being right or wrong. Just read the quotations and then think about them in terms of the concepts, theories, ideas and research results covered in this chapter of the textbook.

"Pardon me, you are not engaged to any one. When you do become engaged to some one, I, or your father, should his health permit him, will inform you of the fact. An engagement should come on a young girl as a surprise, pleasant or unpleasant, as the case may be. It is hardly a matter that she could be allowed to arrange for herself.
Oscar Wilde (1584-1900), Anglo-Irish playwright, author. Lady Bracknell to Gwendolen, in *The Importance of Being Earnest*, act 1.

"If we are a metaphor of the universe, the human couple is the metaphor par excellence, the point of intersection of all forces and the seed of all forms. The couple is time recaptured, the return to the time before time."
Octavio Paz (b. 1914), Mexican poet. *Alternating Current*. "Andre Breton or the Quest of the Beginning" (1967).

"Marry a mountain girl and you marry the whole mountain."
Irish Proverb.

TRUE-FALSE

If you have studied and understand the textbook, you should be able to answer correctly the following true-false questions. Print a T or F in the answer space for each question, and then check your answers with the correct answers at the end of this chapter.

_____ 1. Courtly love flourished during the Middle Ages, involved a great deal of idealization, and did not require the couple to live together.

_____ 2. A "dowry" is an older woman, usually unmarried, who has considerable wealth.

_____ 3. Exchange theory of mate selection assumes that people are largely rational and consider their assets and liabilities as they bargain in the marriage market.

_____ 4. The traditional exchange in marriage is related to gender roles.

_____ 5. Older men and women may more easily evaluate prospective partners because personal and economic characteristics are more established.

_____ 6. The sex ratio refers to the ratio of successful sexual encounters to unsuccessful sexual encounters.

_____ 7. Among blacks, the sex ratio is at or slight above 100.

_____ 8. Currently, androgynous social expectations and behavior have equalized the bargaining positions of men and women in the marriage market.

_____ 9. Karin, age 25, is a college-educated insurance agent and has married a man who is also age 25, and who also a college-educated insurance agent. Karin has behaved according to the marriage gradient.

_____ 10. Karin and her husband, as described in question #9 above, are an excellent example of homogamy.

_____ 11. The "pool of eligibles" refers to the increased number of now-available divorced men in a population.

_____ 12. Propinquity refers to a person's level of susceptibility to romantic or erotic stimuli, or, to the probability of one falling easily in love with the right person.

_____ 13. Fred hates cooking, but he says that while he is living together with Jane, because it seems to make her happy, he does his share of it and try to appear as if he is enjoying it. Later, after they're married, he intends to "back off cooking any more." Fred is "imaging."

_____ 14. "Hypergamy" refers to excess concern or anxiety over not yet being married.

_____ 15. Of all interracial marriages, about sixty percent are between blacks and whites.

_____ 16. Research indicates that heterogamous marriages tend to be more stable than those that are not heterogamous.

_____ 17. SVR refers to a three-stage filtering sequence of mate selection.

_____ 18. A study of college students found that "dating" as a pattern of mate selections has almost completely ceased to exist, having been replaced almost completely by what young adults call "getting together."

_____ 19. The courtship game continues to follow gender-role stereotypes.

_____ 20. Karen and Danny have been cohabiting. Now they have decided to make a lifetime commit to each other and marry, though they never intend to have children. This long-term couple commitment signals their movement into the second stage in Mead's concept of a "two-stage marriage."

_____ 21. Pre-marriage "cohabitation" as a time for testing is an especially appropriate time for "imaging."

_____ 22. Serial cohabitors tend to have higher divorce rates.

_____ 23. About 25 percent of teenage first births are legitimated by marriage before birth.

MULTIPLE CHOICE

If you have studied and understand the textbook, you should be able to answer correctly the following multiple choice questions. Some of the questions are general and some are specific. Print the correct alternative letter-answer in the space to the left of each question. Check your answers with the correct answers at the end of this chapter. If you understand the chapter very well, you should miss few or none of these questions. For each question you miss, review the pertinent textbook material to understand *why* your answer is incorrect.

_____ 1. "Romantic love" is closest in meaning to which of these?
a. poetic bargaining
b. the heraldic code
c. courtly love
d. Napoleonic love

_____ 2. The exchange theory of dating and mate selection is closest to which of these?
a. leveling
b. bargaining
c. imaging
d. emotionality

_____ 3. The sex ratio is calculated in terms of:
a. persons' levels of physical attractiveness
b. a kind of "batting average" regarding sexual conquests
c. the extent to which men and women embody ideals of masculinity and femininity
d. the number of males per 100 females

_____ 4. Hulda, a Swedish-American whose Protestant religion is important to her and to her family, has married Derby, an Irish-American who, interestingly enough, is Jewish and goes to synagogue every week. This illustrates:
a. homogamy
b. inappropriate pool of eligibles
c. exogamy
d. endogamy

_____ 5. Pete, independently wealthy, a bank manager, an influential politician, age 37, and an African American, married Anna, a white bank teller age, 33, who feels politics is very important and appreciates Andy's social position in the community. Anna feels that by marrying Pete she has "come up in the world." To the extent that Anna is correct about that, it illustrates:
a. hypergamy
b. homogamy
c. endogamy
d. propinquity

_____ 6. Of all marriages, the proportion that are interracial is about what percent?
a. less than one percent
b. five percent
c. sixteen percent
d. twenty-one percent

_____ 7. SVR directly reflects aspects of determining:
a. levels of marital satisfaction in the first months of marriage/cohabitation
b. who is initially attracted to whom in the first stage of a couple relationship, in the first moments they meet
c. likelihood of divorce, once marriage has occurred, or, among cohabitors, which cohabitation arrangements are likeliest to end in disagreement
d. none of the above

_____ 8. "Imaging" consists of:

a. presenting an unrealistically attractive one oneself to another person, hoping it will be taken as real, thus attracting and holding the other person
b. trying to keep before one's mind a conception of the kind of person one wants to be, of how one wants to become, hoping in the near future to turn that into reality
c. the process whereby couples troubled relationships try to keep "outsiders" from becoming aware of the troubled couple's difficulties
d. none of the above

_____ 9. Mead's two-stage marriage consists of _____________ marriage and parental marriage.

a. affectionate
b. quasi-
c. individual
d. nontraditional

_____ 10. As traditional courtship progresses, _________ places less emphasis on the end result, marriage, than ___________ does.

a. the man; the woman
b. homogamy; heterogamy
c. interracial dating; interreligious
d. getting together; dating

_____ 11. About half of all courtship relationships continue after there has been:

a. courtship violence
b. a brief period of temporary separation if the separation was unavoidable
c. an emotional agreement as to the nature of the temporary commitment between the partners
d. a mid-stream confusion about where the relationship is going, although an equal proportion of relationship terminate after this phase

_____ 12. About what percent of white women's first cohabiting relationships resulted in marriage?

a. one-quarter
b. one-third
c. slightly more than half
d. two-thirds

_____ 13. Which of these is *not* one of the four common patterns of cohabitation?

a. convenience
b. testing
c. emancipation
d. habituated

_____ 14. When a person cohabits with several persons, one after another, this is termed:
a. time series cohabitation
b. serial cohabitation
c. cohabitation filtering
d. dysfunctional cohabitation

_____ 15. Research tends to suggest that women _________ more readily than do men.
a. approach relationships nonrationally
b. see the partner from unrealistically flattering perspective
c. tend to fall out of love
d. become relationship micro-politicians, manipulating aspects of the relationship

_____ 16. Which of the following is associated with persons marrying early?
a. Roman Catholic (as compared with Protestant) religious background
b. low socioeconomic origins
c. both partners being "balanced" in terms of emotional commit, agreement on important values, and appropriate behavioral norms
d. none of the above

_____ 17. Which of the following has been found to be *most* significantly associated with breakup after marriage has occurred?
a. premarital pregnancy
b. premarital birth
c. disagreement abpit using/not using contraception, and if used, by whom
d. sexual "imaging"

_____ 18. According to the text, for teenage parents, babies:
a. help cement an otherwise shaky relationship
b. often keep the grandparents in contact with each other if/when divorce occurs
c. usually serve as a test of whether or not the couple have a relationship worth keeping
d. can interfere with the young parents achieving their educational and career goals

_____ 19. According to the text, using desirable physical appearance, social pressure, and economic advancement as reasons for marriage are:
a. some things that should be considered before legal union is seriously contemplated
b. associated with unstable marriage relationships
c. topics that should be included in any reputable premarital counseling session
d. openly considered by lower-class partners but only subconsciously considered by partners in the middle-class and/or above

_____ 20. Which of the following is *not* among Knox's positive reasons for marrying?
a. loneliness
b. obligation
c. social pressure
d. all of the above

Your Opinion, Please: Historically, research found little difference in the marriages of those who cohabited before marriage and those who did not. But evidence in recent years seems to suggest that "trial marriage" may have a negative effect on marital success. What do you make of the difference between what researchers have found in the past compared with what recent research has concluded? Which way do you think it "is"? Or do you think that in the past it might have been one way and these days it is another? Whether you think cohabitation increases, decreases, or makes no difference on chances for marital success, what is the research *evidence* for each of these three points of view?

SHORT-ANSWER ESSAY QUESTIONS

The following are sample short-answer essay questions — questions of the type you may be asked if your instructor uses questions like these. Even if your instructor does not use questions like these, you can help organize and consolidate your learning if you can answer these questions in a well-organized and complete manner.

Do not think that it is easier to write adequate answer for brief essays than for longer essays. Short-answer essays may be more challenging because answers are required to be both brief and complete. And after you have answered these questions below, construct some similar questions of your own devising, and then answer them. The Study Questions at the end of each textbook chapter make very good short-answer essay questions.

1. Clearly define both endogamy and homogamy. For each, give an example, and explain or make clear *why* your example is a good example.

2. In what way(s) is cohabitation different from a two-stage marriage?

3. Distinguish between dating and getting together, being careful to be both brief but complete in your distinction(s). Since few things are *all* good or *all* bad, what, for example, are the attractions and drawbacks of each?

4. How does the sex ratio affect advantage and/or disadvantage in obtaining a spouse?

5. Summarize the research findings for one of the following: interreligious marriages, interclass marriages, interracial marriages. (Don't just throw in numbers for numbers' sake, but try to include appropriate important statistics or percents whenever it helps to give credibility to the point(s) you are attempting to make in your answer.

6. To what extent are egalitarian courtship patterns an "improvement" over traditional courtship patterns, according to the textbook?

7. What happens to couple relationships when courtship violence occurs? Be brief, specific, yet complete in your answer.

8. Maria's and her boyfriend Tony, are in their late teens, come from different ethnic/racial backgrounds and religions, and are experiencing a lot of a different religion, have only recently met, have been sexually intimate (she may be pregnant), and they are both experiencing a great deal of pain from the enormous social pressure they face urging them to end the relationship. Still, they both feel deep love and strong multi-level attraction for each other, which they feel will enable them to triumph over all odds and have a successful, happy long-term relationship. Her best friend, Anita, has suggested that the relationship is a bad idea for lots of reasons. Maria replies, "I know, but ..." In the space provided below, what does the textbook suggest about the probability of success for a relationship such as this? (Although you could probably write a longer answer, try to confine your answer to the space below, the purpose being to force you to write only the most essential elements of an answer.)

ESSAY

The following are sample essay questions — questions of the type you may be asked if your instructor uses essay questions. Even if your instructor does not use essay questions, you can help organize and consolidate your learning if you can answer these questions in a well-organized and complete manner. Of these essay questions, the third essay question is usually the most challenging.

1. Re-read short-answer essay question #8 just above. Write a longer, more complete answer in essay form.

2. Explore the extent to which mate selection is a "bargaining" relationship. Support your answer with pertinent data and research findings or research results.

3. Jeff has married Sharon. Sharon is white and Roman Catholic, just like him. They met at a private liberal arts college with a predominantly white, middle- and upper-middle class student population. Jeff was majoring in sociology and Sharon was majoring in Psychology, so they met and got to know one another while taking classes in the same academic building. They fell in love and married. What sociological concepts, factors, and/or theories help us to understand how this could have happened?

4. Imagine that your friend has confided in your that he/she seems to be involved in the process of cohabit and, since you are in a family sociology class, wants to know "the facts" about the effects of cohabitation on a relationship, especially as probability of eventual marriage and happiness in marriage. Since your friend is serious in trying to get some helpful information for responsible decision making, what can you tell her/him that might be helpful? Where possible, make specific reference(s) to research results or findings to help give your friend more confidence in your answer.

ANSWERS

CHAPTER SUMMARY

in the **marrige market**
and **physical attractiveness**
propinquity, social pressure
getting together, and **cohabitation**
who **marry too young** are

COMPLETION

1. courtly love
2. dowry
3. marriage market
4. sex ratio
5. exchange theory
6. pool of eligibles
7. homogamy
8. endogamy
9. hypergamy
10. exogamy
11. heterogamy
12. getting together
13. cohabitation

TRUE-FALSE

1. T
2. F
3. T
4. T
5. T
6. F
7. F
8. F
9. F
10. T
11. F
12. F
13. T
14. F
15. F
16. F
17. T
18. F
19. T
20. F
21. F
22. T
23. T

MULTIPLE CHOICE

1. C
2. B
3. D
4. C
5. A
6. B
7. D
8. A
9. C
10. D

11. A
12. C
13. D
14. B
15. C
16. B
17. B
18. D
19. B
20 D

CHAPTER 8

MARRIAGE: A UNIQUE RELATIONSHIP

This chapter marriage as a unique relationship by explaining a few basic kinship terms, pointing out that the basic fact that each kinship types unfailingly specifies a range of expected obligations and marriage relationships. Five types of marriage relationships are discussed. The Marriage Premise is explored, along with its concomitants of the expectations of permanence and primariness, and the effects of extramarital sex. Same-sex "marriage" is discussed. The chapter closes by examining the choices made throughout life —especially in the first years of marriage—and closes by comparing static versus flexible marriages, and the pro's and con's of marriage agreements or contracts.

CHAPTER SUMMARY

Although marriage is less permanent and more flexible than it has ever been, it is still unique and set apart from other human relationships. Although legal marriage is not possible for same-sex couples in the United States, this situation may soon change. Meanwhile, the marriage premise includes expectations of permanence and ______________________________. As both of these expectations come to depend less on legal definitions and social conventions, partners need to invest more effort in sustaining a marriage.

Because few data are available on the period just before marriage, there is much concern about preparation for marriage—is it adequate in today's society? Premarital counseling and family life education in the schools are two approaches that have been developed, but we need more research data on their effectiveness. We need to learn more about the transition into marriage, although it appears that the early years of marriage tend to be a happy time.

Two opposite poles on a continuum of marriage are the utilitarian marriage and the ___________________________ marriage. Most marriages fall somewhere in between. Some frequently occurring marital types are the ___________________-habituated, the devitalized, the passive-________________________, the vital, and the ______________________ marriage.

Partners change over the course of a marriage, so a relationship needs to be flexible if it is to continue to be intrinsically satisfying. A static marriages are usually a devitalized ______________. Marriage or domestic partner agreements, which can be renegotiated as the need arises, are on useful way of coming to mutual agreement. Working on a marriage agreement together can help partners develop a couple identity—one of the tasks of early marriage.

Point to Ponder: In this chapter you will find out about the many ways in which couples live their lives. Some couples revel in each other's lives and fulfill each other. Other couples seem to be friends, but little else And there are other types of couples as well. Question: What *should* marriage "be"? Should marriage be a deeply fulfilling experience, or is a pleasant friendship enough? Put differently, what is "adjustment" in marriage? Is a spouse well-adjusted if she/he is happy and fulfilled, but the partner is miserable? For some couples, should an uneasy truce be "as good as it gets"?

KEY TERMS

You should be able to explain the concepts listed below. In your explanation, try to avoid using the concept you are explaining. You should be able to *give several examples* of each concept and to *explain why* each example is an example.

consanguineous 220
conjugal 220
dominant dyad 220
family of orientation 220
family of procreation 220
la familia 222
parallel relationship pattern 222
interactional pattern 222
utilitarian marriage 222
intrinsic marriage 222
conflict-habituated marriage 222
devitalized marriage 223
passive-congenial marriage 223
vital marriages 224
total marriages 224
marriage 224
marriage premise 224
primariness 226
sexual exclusivity 226
jealousy 230
role making 236
static (closed) marriages 238
flexible marriage 239
personal marriage agreements 239
relationship agreements 239

COMPLETION

Complete the following sentences by selecting the correct alternative from the Key Terms listed above. Some key terms may be used more than once. Some may not be used at all. Filling in a blank may require more than one word.

1. The family most people grow up in is called their "family of ___________________."

2. One's family of ________________________ is formed by marrying and having children.

3. A ______________________ marriage is intrinsic, and is like a vital marriage, and it is more multifaceted.

4. Partners who rely on formal rules to maintain their marriage, feelings of intimacy and love, and sense of commitment, and use repetitive patterns of unchanging behavior, and go about their relationship in the same way day after day, are involved in a ______________________ marriage.

5. A ______________________ is a marriage that allows and encourages partners to grow and change, and allows for role negotiation as partners' needs change.

6. Every society has a ______________________, or, a centrally important twosome, and at least among white, middle-class Americans, it is expected to take precedence over any others.

7. The expectation by married partners that both will keep each other as the most important person in their lives is called ______________________.

8. ______________________ is the emotional pain, anger, and uncertainty arising when a valued relationship is threatened or perceived to be threatened.

9. A couple's relationship lies somewhere along the continuum between utilitarian marriage and ______________________ marriage.

10. One way for couples to develop flexible marriage is to write ______________________ or ______________________.

11. Family sociologist Jesse Bernard noted what she called a parallel relationship pattern among spouses in the working class, which she distinguished from the ______________________, a pattern more frequently found in middle-class couples.

12. In a ______________________, the partners experience considerable tension and unresolved conflict, quarrel habitually, and are accustomed to this continuing pattern of behavior.

13. When they marry and accept the responsibility to keep each other primary in their lives, and to work hard to ensure that the relationship continues, this is the ______________________.

14. When partners negotiate behavioral expectations rather than mutely accept standard or normative patterns of behavior, this is termed ______________________.

KEY RESEARCH STUDIES

You should be familiar with the main question being investigated and the research findings or research for the following studies:

Cuber and Harroff: research about utilitarian and intrinsic marriage
Cuber and Harroff: research about five kinds of marital relationships
Masters, Johnson, and Kolodny: research about extramarital affairs

Quotations to assess: Read the following quotations and then think about them, assessing or evaluating their meanings or evaluating them in terms of the text material in this chapter. Do not worry about being right or wrong. Just read the quotations and then think about them in terms of the concepts, theories, ideas and research results covered in this chapter of the textbook.

"The best friend is likely to acquire the best wife, because a good marriage is based on the talent for friendship."
Friedrich Nietzsche (1875-1900), German philosopher. *Human, All Too Human*, ch. 7, aph. 378.

"Set me as a seal upon thine heart, as a seal upon thine arm; for love is strong as death; Jealousy is cruel as the grave; the coals thereof are coals of fire, which hath a most vehement flame."
Hebrew Bible *Song of Solomon* 8:6

"Before marriage a man will go home and lie awake all night thinking about something you said; after marriage, he'll go to sleep before you finish saying it."
Helen Rowland (1875-1950), U.S. journalist. *A Guide to Men*, "First Interlude" (1922).

"Marriage is an act of will that signifies and involves a mutual gift, which unites the spouses and binds them to their eventual souls, with whom they make up a sole family—a domestic church."
Pope John Paul I [Karol Wojtyla] (b. 1920), Polish ecclesiastic, pope. Quoted in: *Observer* (London, 31 Jan. 1982).

"Jealousy is all the fun you *think* they had ..."
Erica Jong (b. 1942), U.S. author. *How To Save Your Own Life*, "Bennett tells all in Woodstock...," epigraph (1977).

TRUE-FALSE

If you have studied and understand the textbook, you should be able to answer correctly the following true-false questions. Print a T or F in the answer space for each question, and then check your answers with the correct answers at the end of this chapter.

_____ 1. The marriage premise refers to the assumption that marriage is no longer "until death parts us."

_____ 2. Flexible marriages anticipate that one or both of the partners will change in time, and encourages or welcome such change.

_____ 3. The effect an extramarital affair has on a marriage is not always adverse or negative only.

_____ 4. A utilitarian marriage is one begun or maintained for primarily practical purposes.

_____ 5. Intrinsic marriages are less vulnerable to divorce than are utilitarian marriages.

_____ 6. Conflict-habituated marriages almost without exception end in divorce.

_____ 7. Emotional emptiness does not necessarily threaten the stability of a marriage.

_____ 8. Of the five types of marriage discussed by Cuber and Harroff, the most all-encompassing is the "vital" marriage.

_____ 9. Partners in static marriages rely on their formal, legal bond to enforce permanence.

_____ 10. Negotiating personal contracts helps to intensify the romanticism that is so important in forming sound engagements and lasting marriages.

_____ 11. Ned and Karin have married. For each partner, this new pair-bond is now the family of orientation.

_____ 12. Hazel and Orin have a lot of conflict in their marriage. They have always had a lot of conflict. Each expects that the conflict will continue. Almost certainly, their marriage will end in divorce if one or the other of them does not die prematurely.

_____ 13. Alvin and Debbie share a strong commitment to each other, but their agreement is that although each may at some time have sex with someone else, that they will be involved in no pregnancy or childbearing with anyone else outside their pair-bond. This illustrates what the textbook calls "sexual exclusivity."

_____ 14. Long-term affairs tend to be more complex than short-term affairs.

_____ 15. The state of Hawaii has been active in considering the legal status of same-sex marriages.

MULTIPLE CHOICE

If you have studied and understand the textbook, you should be able to answer correctly the following multiple choice questions. Some of the questions are general and some are specific. Print the correct alternative letter-answer in the space to the left of each question. Check your answers with the correct answers at the end of this chapter. If you understand the chapter very well, you should miss few or none of these questions. For each question you miss, review the pertinent textbook material to understand *why* your answer is incorrect.

_____ 1. Rodolpho and Mimi are determined to put each other first in their lives, to try to make each other happy, and to try to make their relationship endure. Which of the following applies to this example?
a. aggressive-conformist marriage
b. the marriage premise
c. the marriage gradient
d. the N.A.S.H. process

_____ 2. The commitment of both partners to keeping each other the most important person in their lives is called:
a. fronting
b. Christian love
c. humanistic love or "agape"
d. primariness

_____ 3. Newlyweds negotiate expectations for sex and intimacy, establish communication and decision-making patterns, and come to some agreements about childbearing. According to the textbook, such newlyweds are involved in:
a. creative institutionalization
b. role making
c. boundary-maintaining
d. primary institutionalization

_____ 4. Which of the following is most incompatible with "sexual exclusivity"?
a. anticipatory socialization
b. swinging
c. professionalization
d. social mobility

_____ 5. The "intrinsic marriage" offers:
a. increased promise of upward social mobility
b. community rewards
c. external rewards, because each is free to see rewards (other than sexual) outside the relationship
d. intense emotional rewards

_____ 6. A "vital marriage" is one in which the partners:
a. are brought together because of some pressing mutual crisis that they solve jointly
b. marry because one of them has a pressing need or desire for marriage
c. marry because someone else wants the marriage to happen
d. none of the above

_____ 7. "Total" marriages are also:
a. superficial
b. intense but tend to be more divorce-plagued
c. on the increase
d. intrinsic

_____ 8. According to the text, the nature and quality of a marital relationship has a great deal to do with:
a. the political circumstances in which marital life must be lived
b. the economic circumstances in which life must be lived
c. the choices partners make
d. the personal wishes of the individual partners

_____ 9. A flexible marriage is one in which:
a. a marriage contract or personal contract specifies what the requirements are
b. people may change in terms of their sexual preferences
c. the partners are not subject to the ordinary legal system and its definitions
d. partners are relatively free to grow and change

_____ 10. Relationship agreements differ from marriage contracts in that relationship agreements:
a. specify emotional outcomes only
b. specify practical outcomes only
c. have an impact on emotional life, whereas marriage contracts have implications for behavior
d. can apply to both married and unmarried couples

_____ 11. One important reason for writing a marriage agreement is that it helps partners to be aware of and to avoid:
a. ending up in closed marriages that are symbiotic
b. choosing closed marriage by default
c. the pitfalls of traditionalism in family and childbearing
d. exorbitant attorney's fees associated with divorce and dissolution of marriage

_____ 12. Which of the following is explicitly mentioned by the text as an important reason for writing a marriage agreement?
a. being specific about sex role expectations
b. discovering the partner's past affectional attachments
c. uncovering emotional disappointments that may affect the current relationship
d. reducing the possibility of a failed relationship by 70-85 percent

_____ 13. Consanguineous relatives are those who:
a. are "blood relatives"—persons to whom one is related by biological heredity
b. are fictive kin—*referred to* by kinship terms, but to whom there is no relationship by blood or marriage
c. become relatives through marriage, common law marriage, or marriage-like cohabitation
d. become relatives through adoption and are thus "real" relatives, but who are not "blood" relatives

_____ 14. Mothers-in-law, fathers-in-law, sisters-in-law and the like are ___________ relationships, acquired through marriage.
a. fictive
b. conjugal
c. quasi-legal
d. conflict-habituated

_____ 15. Patricia and Jeff met the first of classes at the university and discovered that they complemented one another almost perfectly. Jeff couldn't cook; Patricia could. Patricia lacked any sense of how to repair even so much as a blown fuse; Jeff was very handy with household repair tasks and tools. They found they also got along very well in terms of affection and sex, meeting each other's needs in that sense. Both felt that living together made practical sense, but agreed that doings so without being married was a sin. They married. Their marriage is an example of:
a. temporary advantage
b. practical-stable marriage
c. utilitarian
d. none of the above

_____ 16. Betty and Alvaro both wanted a vibrant, intensely emotional marriage. For awhile, they had one. Then the "zing" went out of their relationship and, though they continue their relationship with an absence of conflict, the vibrancy has gone from the relationship. Theirs is a(n):
a. devitalized marriage
b. passive-congenial marriage
c. vital marriage
d. none of the above

_____ 17. Since 1974, there has been an increase in Americans who think that:
a. cohabitation is always wrong
b. marriage agreements help bring stability and clarity to marriage relationships
c. same-sex relationships and sexual orientation are set at birth and cannot be changed
d. extramarital sex is always wrong

_____ 18. _________ marital relationships are likely to be utilitarian.
a. working-class
b. middle-class
c. African-American
d. Hispanic

_____ 19. Hedonistic affairs rarely lead to:
a. emotional entanglements
b. STD's
c. genitally-oriented relationships
d. any of the above

_____ 20. Jim, who has always been anxious about growing older, finds it traumatic that the last of the family's children—now age 24—has finally left home for separate lodgings. Jim is frantic about this additional sign of his own aging, and largely because of it gets involved in a very temporary:
a. intimacy reduction affair
b. reality-reducing affair
c. reactive affair
d. marriage maintenance affair

_____ 21. By a(n) _________, marriage is defined as a union between one man and one woman.
a. *amicus* court brief (1924)
b. U.S. Conference of Governors decision (1924)
c. constitutional amendment (1937)
d. U.S. Supreme Court decision (1974)

_____ 22. The State of_________ Supreme Court voted _____ that refusal to recognize same-sex marriages may violate sex discrimination laws.
a. California; 7 to 3
b. Hawaii; 3 to 2
c. California; 2 to 10
d. Hawaii; 7 to 3

_____ 23. People will _________, even if there are promises to the contrary.
a. have extra-marital affairs
b. tend to embody regression toward the emotional "average," moving away from being especially joyous and/or especially sad
c. change
d. none of the above

Your Opinion, Please: Some say that keeping a marriage not only alive and vibrant takes *work*. Do you think it's true? What about the idea that if a person can't be his or her "natural" or authentic self—whatever that is—and be happily married, then divorce is justified. Others have different ideas. How much do *you* think that a person ought to have to *work* at a marriage?

SHORT-ANSWER ESSAY QUESTIONS

The following are sample short-answer essay questions — questions of the type you may be asked if your instructor uses questions like these. Even if your instructor does not use questions like these, you can help organize and consolidate your learning if you can answer these questions in a well-organized and complete manner.

Do not think that it is easier to write adequate answer for brief essays than for longer essays. Short-answer essays may be more challenging because answers are required to be both brief and complete. And after you have answered these questions below, construct some similar questions of your own devising, and then answer them. The Study Questions at the end of each textbook chapter make very good short-answer essay questions.

1. Briefly but completely distinguish between utilitarian marriage and intrinsic marriage.

2. Taking care to point out similarities and differences, compare devitalized marriage and passive-congenial marriage.

3. Of "vital marriage" and "total marriage," one of the two is mentioned by the text as having more potentially negative consequences. Which has more potentially negative consequences, and what are those negative consequences?

4. Distinguish between intimacy reduction affairs and reactive affairs, and give an illustration of either one, making sure that it is clear from your example why it is a good example.

ESSAY

The following are sample essay questions — questions of the type you may be asked if your instructor uses essay questions. Even if your instructor does not use essay questions, you can help organize and consolidate your learning if you can answer these questions in a well-organized and complete manner. Of these essay questions, the third essay question is usually the most challenging.

1. What are the view of Gay Rights activists on same-sex marriage?

2. Some people feel about a personal marriage agreement or a relationship agreement as many others feel about a prenuptial agreement: If you can't just trust the other person to do the right thing, then the proposed marriage is probably a bad idea. What position does the textbook seem to take on this question as regards personal marriage agreements and/or relationship agreements?

3. The textbook states "By getting married, partners accept the responsibility to keep each other primary in their lives and to work hard to ensure that their relationship continues. Essentially, this is the marriage premise." *Permanence* and *primariness.*

   Question: What are the merits, if any, to introjecting a conditional element into the marriage premise, so that from the outset of the marriage, the understanding is that the understanding is that the marriage is "for now" but is not necessarily permanent, that the partner is given primariness now, but that this may not continue indefinitely? Discuss, using a combination of your own ideas and ideas from the textbook.

ANSWERS

CHAPTER SUMMARY

and **primariness**
and the **intrinsic** marriage
conflict-habituated
passive-**congenial**
and the **total** marriage
usually a **devitalized** marriage

COMPLETION

1. orientation
2. procreation
3. total
4. static (closed) marriage
5. flexible marriage
6. dominant dyad
7. primariness
8. jealousy
9. intrinsic
10. personal marriage agreements (or) relationship agreements
11. interactional pattern
12. conflict-habituated marriage
13. marriage premise
14. role making

TRUE-FALSE

1. F
2. T
3. T
4. T
5. F
6. F
7. T
8. F
9. T
10. F
11. F
12. F
13. F
14. T
15. T

MULTIPLE CHOICE

1. B
2. D
3. B
4. B
5. D
6. D
7. D
8. C
9. D
10. D
11. B
12. A
13. A
14. B
15. C
16. A
17. D
18. A
19. A
20. C
21. D
22. B
23. C

CHAPTER 9

COMMUNICATION AND CONFLICT RESOLUTION IN MARRIAGES AND FAMILIES

This chapter focuses on couple communication in general, especially communicating about conflicts. Much unhappiness among couples sometimes results from partners' attempts to deny or ignore conflict. Denying the existing of conflict or attempting to cope with it in unproductive ways can have a variety of negative consequences that accrue to deny or ignoring conflict. The chapter reviews some practices that couples should avoid when fighting because such practices seem to worsen rather than improve matters. The chapter ends by discussing some healthy attitudes and proposing some guidelines for communicating constructively, and examines some differences among marriages in their preferred communication styles.

CHAPTER SUMMARY

Families are powerful sources of support for individuals, and they reinforce members' sense of identity. Because the family is powerful, it can cause individuals to feel constrained, "held back," smothered, and over-controlled by others, as well as the negative emotions that accompany such a state of affairs. Tactics such as gaslighting, __________________________, or negative use of the __________________ self can all be stressful or denigrating to an individual.

Even though such family interaction tactics can reach the point of pathology, family conflict itself is an inevitable part of normal family life. Americans are socialized to respect a _________________ taboo, but both sociologists and counselors are recognizing that to deny conflict can be destructive to both individuals and relationships.

Although fighting is a normal part of the most loving relationships, there are right and wrong ways of fighting. Alienating practices should be avoided; they hurt a relationship because they lower partners' self-esteem. Bonding fights, in contrast, can often resolve issues and also bring partners closer together by improving communication. Three important bonding techniques are leveling, using ___- statements, and giving _______________________. In bonding fights both partners win.

Research on marital communication indicates the importance to marriage of both positive communication and the avoidance of a spiral of negativity. Stinnett's research on "strong families" suggests that what makes families cohesive are expressing appreciation for each other, doing things together, having positive communication patterns, being committed to the group, having some spiritual orientation, and being able to deal creatively with crises.

Partners need to be attentive to typical gender differences in communication and to other cultural differences as well. There is some variation in the patterns of marital communication that work for couples, depending on their model of marriage. Traditionals, independents and _____________ have different ideologies of marriage, prefer different levels of interdependence, and are comfortable with different levels of overt conflict.

> **Point to Ponder**: A person might say "Yes, I'd like to use some of those fair-fighting techniques, some of those more effective ways of facing and resolving conflicts. But I think that even if I use "I-statements" or try leveling, I'm likely to get yelled at, get told something I don't want to hear or be given alternatives I don't want to be have to choose between—or something worse. So I usually just ...try to put up with it the best I can and go along with whatever way my partner wants to play it."
>
> What do *you* think about a situation like that? What do you think the authors of the textbook would suggest to a person making a statement such as the one above?

KEY TERMS

You should be able to explain the concepts listed below. In your explanation, try to avoid using the concept you are explaining. You should be able to *give several examples* of each concept and to *explain why* each example is an example.

looking glass self 247
attribution 247
consensual validation 247
gaslighting 248
scapegoating 248
conflict taboo 248
anger "insteads"
passive-aggression 249
sabotage 251
displacement 251
suppression of anger 251
alienating fight tactics 251
gunnysacking 252
kitchen-sink fight 252
mixed, or double, messages 254
bonding fighting 254
leveling 255
feedback 256
checking-it-out 256
report talk 260
rapport talk 260
traditionals 261
separates 263
independents 263
family cohesion 263

COMPLETION

Complete the following sentences by selecting the correct alternative from the Key Terms listed on the previous page. Some key terms may be used more than once. Some may not be used at all. Filling in a blank may require more than one word.

1. The idea that conflict and anger are morally wrong and should be discouraged within the family is called the ______________________.

2. Person whose opinions about each other are very important to each individual's self-esteem are referred to by sociologists as ______________________. Another way of expressing this is to say that these are persons whose standards are important to us as we evaluate our behavior.

3. When a person does not directly express anger to another but instead expresses it indirectly — often through a third person— then ______________________ are being used.

4. ______________________ fighting is the kind of fighting that brings people closer together rather than driving them farther apart.

5. The ______________________ refers to the process by which people come to see themselves as others see them.

6. ______________________ refers to those tactics that increase tension and conflict rather than reduce it.

7. When we assume that persons have certain character traits, we are engaging in the __________ process.

8. Paul is furious with Denise, but "nice husbands" don't slap their wives, so Paul kicks the cat. This can be interpreted as an example of ______________________.

9. A ______________________ fight is one in which other topics, previous unsettled disputes, and things that are "beside the point" to the issue being argued about are thrown into the argument, muddying the water and producing unhelpful results.

10. The idea of ______________________ refers to letting the other person know explicitly and completely exactly what you are feeling and thinking.

11. Comments such as "We're just not interested in making love anymore" or "No, I'd rather go bowling; therefore I can't watch that program with you and your mother" can be interpreted as ______________________________________.

12. In ______________________________ members of each group can come to some agreement about how each person sees the world. They may even come to some agreement about what reality is.

13. Andy plans a lawn party and asks Denise to arrange for the lawn tents. Denise "forgets" to do so and the lawn party is less than a success. Andy is furious with Denise (which is what she perhaps unconsciously wanted and in which she now for the moment gives her delicious delight). This can be interpreted as an example of what the text terms ________________________.

14. To send a ______________________________ is to send a message that is contains internal contradictions.

15. When one partner tries to change or distort the other partner's self-concept, the process is referred to as ________________________.

KEY RESEARCH STUDIES

You should be familiar with the main question being investigated and the research findings for the studies listed below. For each, what question were the researchers trying to answer and what was found when the research results were examined? You should understand the question being asked, know what the researchers found, and be able to answer both general and specific questions about the material.

Fitzpatrick: variation among married couples regarding four marital types of communication
Stinnett: research about six qualities of family strengths

KEY THEORETICAL PERSPECTIVES

The purpose of any theory is to help us to see the world more clearly and explain *why* things are they are, thus leading us to greater understanding. You should be familiar with the following, being able to explain it briefly or at length, give examples, and answer pertinent questions.

C.H. Cooley: the looking glass self
Weigert and Hastings: attribution

Quotations to assess: Read the following quotations and then think about them, assessing or evaluating them in terms of the text material in this chapter. Do not worry about being right or wrong. Just read the quotation and then think about in terms of the concepts, theories, ideas, and research results covered in this chapter of the textbook.

"That is the whole secret of successful fighting. Get your enemy at a disadvantage; and never, on any account, fight him on equal terms."
George Bernard Shaw (1856-1950), Anglo-Irish playwright, critic. Gergius, in *Arms and the Man*, act 2.

"Let not the sun go down upon your wrath."
Bible: New Testament. Ephesians 4:26. St. Paul speaks.

"I love to see a young girl go out and grab the world by the lapels. Life's a bitch. You've got to go out and kick ass."
Maya Angelou (b. 1928), U.S. author, "Kicking Ass," interview in *Girl About Town* (13 Oct. 1986; repr. in *Conversations with Maya Angelou*, 1969).

"Rage cannot be hidden, it can only be dissembled. This dissembling deludes the thoughtless, and strengthens rage and adds, to rage, contempt."
James Baldwin (1924-87), U.S. author. "Stranger in the Village," in *Harper's* (New York, Oct. 1953; repr. in *Notes of a Native Son*, 1955).

"The Anger that appears to be building up between the sexes becomes more virulent with every day that passes. And far from women taking the blame...the fact is that men are invariably portrayed as the bad guys. Being a good man is like being a good Nazi."
David Thomas (b. 1959), British editor, "Fifth Column," Oct. 1991, BBC2. Quoted in *Independent on Sunday* (Londay, 22 March 1992).

TRUE-FALSE

If you have studied and understand the textbook, you should be able to answer correctly the following true-false questions. Some of the questions are general and some are specific. Each is either obviously true or obviously false. None of the questions are tricky. Answer all of the questions by printing a T or F in the answer space for each question, and then check your answers with the correct answers at the end of this chapter. If you find that you miss questions, review the textbook material to discover *why* your answer is incorrect

_____ 1. For virtually all members, family living involves striking a delicate balance between belonging and feeling constrained. Striking such a balance involves effective communication, among other things.

_____ 2. The emotional tone not only of fighting, but of everyday communication, is important.

_____ 3. When attribution is "working," family members exercise a kind of power over a family member and so get her/him not only to *do* but also to *be* what the other family members want.

_____ 4. Consensual validation takes place among families in their own homes and among themselves as they go together into the community. It tends not to take place in more impersonal environment such as elementary schools and colleges.

_____ 5. Gaslighting is a communication process that has as its end result viewing one's partner in an overly romanticized manner.

_____ 6. Nanette is angry at Zeke, but she can't bring herself to express it directly. Instead, she has developed a habit of saying she will do things for him when he asks her to, but then she finds some way to escape doing those things, so his requests go unfulfilled. This is an example of passive-aggression.

_____ 7. Zoe takes the indirect approach when she's angry with her boyfriend. If he wants to go to the zoo, she says zoos bore her. If he wants to go for a bike ride with her, she says its a useless form of exercise. If he wants to spend more time at his new computer, she reminds him intermittently that watching the monitor is for a long period of time is rumored to be bad for the eyes. This is an example of displacement.

_____ 8. According to the textbook, gunnysacking is an undesirable way to fight, whereas kitchen-sink fighting is a more effective way to fight because it is friendlier.

_____ 9. This is an example of a clear message, not a "mixed message": "I like your new dress. You look very good in it, all things considered."

_____ 10. "Feedback" is unnecessary, because there is no need to return insult for insult. Usually it accomplishes little to strike back just because someone has struck out at you.

_____ 11. "Rapport talk" is talk that occurs toward the end of a successfully completed bonding fight, or before an unsuccessfully completed one.

_____ 12. All fighting involves some degree of frustration and hurt feelings.

_____ 13. Fair fighting can sometimes remove barriers that allow a couple to divorce.

_____ 14. The authors of the textbook conclude that Tolstoy was correct when he wrote that happy marriages are all alike, but unhappy marriages are unhappy in their own special ways.

_____ 15. In terms of likelihood of marital happiness, it is less important which ideology or style of marital communication a couple has and more important that they share the same ideology or style, whichever of the several styles that might be.

_____ 16. Fitzpatrick found that "traditionals" scored highest on the Dyadic Adjustment Scale.

_____ 17. Stinnett found that strong families were characterized by several specific characteristics, but "spiritual orientation" was not among those characteristics.

_____ 18. Tactics such as scapegoating can be demeaning, denigrating, or hurtful to the individual on the "receiving end" of the behavior, but using the looking glass is a healthier tactic and usually has positive rather than negative results.

MULTIPLE CHOICE

If you have studied and understand the textbook, you should be able to answer correctly the following multiple choice questions. Some of the questions are general and some are specific. Print the correct alternative letter-answer in the space to the left of each question. Check your answers with the correct answers at the end of this chapter. If you understand the chapter very well, you should miss few or none of these questions. For each question you miss, review the pertinent textbook material to understand *why* your answer is incorrect.

_____ 1. It is very important for people in relationships to know how to fight, and knowing how to fight:
   a. is signaled by whether or not one is able to emerge as a winner in a conflict encounter
   b. is something people usually know how to do only with persons of their own sex or gender, disadvantaging them for couple-fighting in heterosexual relationships
   c. is a skill that "ages well" for people who have it, but that is difficult to learn for people who do not have it
   d. is knowing how to communicate about conflict, a skill not everybody has in equal amount

_____ 2. The looking glass self is:
   a. the process whereby people come to see themselves as others see them
   b. a narcissistic phenomenon in which people have an unrealistic view of themselves
   c. a process whereby people look at themselves and have a conflict about the direction in which they should go
   d. the self one sees in the mirror "the morning after" a serious fight with the partner, or, indeed, on any occasion where self-assessment occurs (also called "the Person in the Mirror")

_____ 3. Attribution is:
   a. the ascribing of certain character traits to persons
   b. feeling that a man or woman has many of the attributes that society says make a person socially attractive
   c. the process whereby some reasons are found for family events that need explaining so that temporary sense can be made from them
   d. one of the six methods mentioned by Martinelli as "fair fighting techniques"

_____ 4. Family members' viewpoints are an essential part of:
   a. the kinship negotiation process
   b. determining the rules by which we must live
   c. consensual validation
   d. determining which fighting technique a family uses according to the six possibilities in the Martinelli typology of family fighting

_____ 5. Focusing on one family member to blame for almost everything that goes wrong in a family is what is meant by:
   a. gaslighting
   b. passive-aggression
   c. mixed message
   d. scapegoating

_____ 6. "Gaslighting is best exemplified by which of the following remarks from a husband to his wife?
   a. "I personally prefer the natural light of candles to the artificial light from bulbs."
   b. "Every time you balance the checkbook, you make math errors. Our accountant says the errors are in *your* handwriting. Ask her if you don't believe me."
   c. "Who turned the television on? *You* turned it on. You must be losing your mind."
   d. "Don't turn the ceiling lights on. It's much more pleasant with the light coming from the fireplace. You look even more attractive by natural light."

_____ 7. Gaslighting and scapegoating both illustrate:
   a. methods of communication that have the potential for many error messages
   b. illustrate the negative power of the family to shape self-concepts and behavior
   c. show how individuals can maintain control of situations that threaten to get out of control, if only they know how to manage or shape their verbalizations and behaviors
   d. none of the above

_____ 8. Conflict taboo refers to which of these?
   a. jealous arguments started as a result of jealousies or sexual rivalries
   b. the idea that taboos create conflicts between partners if the taboos are not openly discussed
   c. the concept that close kin become political adversaries when there is family conflict
   d. none of these

_____ 9. According to the textbook, anger "insteads":

a. are a list of conflict techniques that can be used to take the place of angry confrontations and tend to produce healthier outcomes
b. are derived from behavior modification, and are part of family fighting techniques often taught in family therapy settings
c. may be more socially acceptable, but they can be self-destructive kinds of behavior
d. none of the above

_____ 10. Passive-aggression is a concept that refers to:

a. the feeling that one's spouse is being both passive and aggressive, simultaneously
b. the view that one should be more passive in the face of aggression
c. the view that one should not be passive in the face of aggression
d. none of the above

_____ 11. Turning sullen and refusing to talk, or stating "I can't take you seriously when you act this way," are examples of:

a. marriage inflexibility
b. indirect denial of conflict
c. fight evading
d. none of the above

_____ 12. In fights, a rule to use in avoiding attacking a person is to use _____ instead of _____ .

a. "sometimes"; "always," because this allows for some latitude and flexibility
b. gender-neutral words; gender-specific
c. words; actions
d. I-statements; defining reality *for* the other person

_____ 13. A "kitchen-sink" fight" is one in which:

a. the combatants use any argument available to them, whether or not is related to the original purpose of the fight
b. the couple fights at breakfast
c. the couple fights in the kitchen, regardless of the time of day, and where they have at their disposal potentially deadly weapons—knives—that account for so many domestic homicides
d. the "dirtiest dishes" in the couple's history are the topic for discussion during a family argument

_____ 14. Which of these is a method recommended by the text to avoid attacks on a partner's self-esteem?

a. Use "I-statements."
b. Avoid topics in which the partner's self-esteem is an issue.
c. Avoid topics that could have the effect of damaging the partner's self-esteem.
d. Take as much of the blame on oneself as is possible; it tends to lower the partner's tension level and make talk more low-key, enhancing true communication

_____ 15. Feedback exists when one of the partners:
   a. delivers a reply to the partner that is as devastating as the one received from the partner
   b. repeats the partner's criticism but applies it to the partner instead of to oneself
   c. repeats in one's own words what the other partner has said or revealed
   d. uses the present argument to recall and drawn on previous arguments one has won

_____ 16. Which of the following is one of the guidelines for bonding fights?
   a. Don't try to win.
   b. Avoid giving negative information to the partner.
   c. Try to present yourself as you would like to be, not as you think you are.
   d. Fights should always end with an agreement between partners; or, "It's not over till it's over."

_____ 17. The key to _________ is for partners to try to build up, not tear down, each other's self-esteem.
   a. bonding fighting
   b. egoistic fighting
   c. creative fighting
   d. enhancing fighting

_____ 18. A researcher found that _________ tend to rebel against their parents' conflict style; however, it was also found that _______ are more apt to follow their parents' lead when it comes to fighting style.
   a. young men; young women
   b. non-Hispanic whites; African-Americans
   c. African-Americans; non-Hispanic whites
   d. working-class couples; middle-class couples

_____ 19. A researcher found that which of the following tended to marry women very different in conflict style from their own mothers?
   a. men from Irish, Scotch, Scotch-Irish, or English backgrounds
   b. young men
   c. 2nd and 3rd generation immigrants
   d. none of the above

_____ 20. According to the text, men tend to engage in ___ while women tend to engage in ____.
   a. leisure talk; interaction talk
   b. report talk; rapport talk
   c. empty conversation; conversation without listeners
   d. none of the above

Your Opinion, Please: Some people feel that there are times when you are so angry that it is best to keep your thoughts and feelings to yourself. When some things are said, they can never be completely taken back, the slate cannot be wiped clean, and the relationship can never be the same again. Others think that you are entitled to your feelings an beliefs, and that you should let the other person know what you think or feel about things, no matter what the consequences.

Questions: Has someone ever asked you for your honest opinion and, when you gave it, attacked you for your opinion? If it happened, how did it make you feel about the situation at the time? How did it make you feel about wanting more honest, open "leveling" communication with that person?

SHORT-ANSWER ESSAY QUESTIONS

The following are sample short-answer essay questions — questions of the type you may be asked if your instructor uses questions like these. Even if your instructor does not use questions like these, you can help organize and consolidate your learning if you can answer these questions in a well-organized and complete manner.

Do not think that it is easier to write adequate answer for brief essays than for longer essays. Short-answer essays may be more challenging because answers are required to be both brief and complete. And after you have answered these questions below, construct some similar questions of your own devising, and then answer them. The Study Questions at the end of each textbook chapter make very good short-answer essay questions.

1. Briefly explain each of the following: anger "insteads" and "kitchen-sink" fighting.

2. Explain the difference between each of the following: "I-statements" and "leveling"

3. Distinguish between "report talk" and "rapport talk" as styles of communication.

ESSAY

The following are sample essay questions — questions of the type you may be asked if your instructor uses essay questions. Even if your instructor does not use essay questions, you can help organize and consolidate your learning if you can answer these questions in a well-organized and complete manner. Of these essay questions, the third essay question is usually the most challenging.

1. What is the connection between communication in families and Cooley's concept of the looking glass self?

2. Explain the various alienating practices explored in the text. For each, given an example of your own devising, and in your answer make it clear *why* each example is a good example.

3. Explore the origins and consequences of conflict in families. Where appropriate, back up your essay with appropriate specific concepts and research results.

ANSWERS

CHAPTER SUMMARY

looking glass self
scapegoating, or use of
conflict taboo
I-statements
giving **feedback**
separates have

COMPLETION

1. conflict taboo
2. significant others
3. passive-aggression
4. bonding
5. looking-glass self
6. alienating fight tactics
7. attribution
8. displacement
9. kitchen sink
10. leveling
11. anger "insteads"
12. consensual validation
13. sabotage
14. mixed or double message
15. gaslighting

TRUE-FALSE

1. T
2. T
3. T
4. F
5. F
6. T
7. T
8. F
9. F
10. F
11. F
12. T
13. T
14. F
15. T
16. T
17. F
18. F

MULTIPLE CHOICE

1. A
2. B
3. A
4. C
5. D
6. C
7. B
8. D
9. C
10. D
11. D
12. D
13. A
14 A
15. C
16. A
17. A
18. A
19. B
20. B

CHAPTER 10

POWER AND VIOLENCE IN MARRIAGES AND FAMILIES

This chapter explores what power is, how it is distributed in families, and the violence that sometimes accompanies the exercise of power. Family diversity can be seen in conjugal power and decision making, with differences based on resources, gender, cultural context, social class, as well as racial/ethnic diversity. Although some couples play power politics in a number of ways, there are alternatives, including the change to a no-power relationship. Family abuse is reviewed, including wife abuse, husband abuse, abuse among lesbian and gay men couples, and elderly abuse. The chapter ends by reviewing basic rights in any relationship.

CHAPTER SUMMARY

________________, the ability to exercise one's will, may rest on cultural authority, economic and personal resources that are gender based, love and emotional dependence, interpersonal manipulation, or physical violence.

The relative power of a husband and wife in a marriage varies by national background and race, religion, and class. It varies by whether or not the wife works and with the presence and age of children. American marriages experience a tension between male dominance and egalitarianism. Studies of married couples, cohabiting couples, and gay and lesbian couples illustrate the significance of economic-based power, as well as the possibility for couples to consciously work toward more egalitarian relationships.

Physical violence is most commonly used in the absence of other resources. Although men and women are equally likely to abuse their spouses, the circumstances and outcomes of marital violence indicate that wife abuse is a more crucial social problem. It is has received the most programmatic attention. Recently, programs have been developed for male abusers, but less attention has been paid to male victims. Studies indicating that arrest is an important deterrent to future wife abuse illustrate the importance of public policies that meet family needs.

Economic hardships and concerns (among parents of all social classes and races) can lead to physical and/or emotional____________________________ —a serious problem in our society and probably far more common that statistics indicate. One difficulty is drawing a clear distinction between "normal" child rearing and ___________________.

______________________________ and neglect is a new area of research, but initial data suggest that abused elderly are often financially independent and abused most frequently by dependent adult children or by elderly spouses. Although some scholars view elder abuse and neglect as primarily a caregiving issue, this chapter presents elder abuse as a form of family violence and as a family power issue.

Point to Ponder: It appears that the partner with the more resources tends to have the most power in a relationship. This being true, do you think that a married person's power will change vis-à-vis the spouse if the spouse is able to complete a college degree? Why do you come to that conclusion? What difference does it make, do you think, if the partners really love each other? Does love triumph over resources? Do resources win out over love? Or what?

KEY TERMS

You should be able to explain the concepts listed below. In your explanation, try to avoid using the concept you are explaining. You should be able to *give several examples* of each concept and to *explain why* each example is an example.

power 270
personal power 270
social power 270
conjugal power 271
coercive power 271
reward power 271
expert power 271
informational power 271
referent power 271
legitimate power 271
resource hypothesis 271
principle of least interest 277
relative love and need theory 277
no power 280
power politics 280
neutralizing power 282
violence between intimates 286
marital rape 289
three-phase cycle of violence 289
child abuse 295
sexual abuse 296
child neglect 296
emotional child abuse or neglect 296
elder abuse 296
elder neglect 296

COMPLETION

Complete the following sentences by selecting the correct alternative from the Key Terms listed above. Some key terms may be used more than once. Some may not be used at all. Filling in a blank may require more than one word.

1. ____________________________ can be defined as the ability to exercise one's will (or, to get one's way.

2. In a __________________________ situation, each partner knows vicariously how to play the other's role, that is, knows how to play either the dominant or the submissive role.

3. The ______________________________ refers to the fact that the partner who has less to lose from ending the relationship is the one who is more pat to exploit the other partner.

4. When a subordinate weakens the powerful person's control by refusing to cooperate in that power, the term ______________________________ is applicable.

5. Power exercised over oneself is called ______________________________.

6. Blood and Wolfe in their ______________________________ , argued that power in marriage comes mainly from the adequacy of the various things one can draw on or "fall back on" as leverage to "get one's way."

7. Power between married persons is termed ______________________________.

8. ______________________________ means that both partners wield about equal power.

9. In ______________________________, a subordinate weakens the powerful persons control by refusing to cooperate in that power, not to gain dominion or control but instead to move toward an equal position.

10. ______________________________ is the kind of power that is based on the persuasive content of what the dominant person tells another individual, as when a patient is convinced by the information received from a trusted physician, or when a parent listens to an qualified substance abuse counselor talk about drug substances currently being abused by teens.

KEY RESEARCH STUDIES

You should be familiar with the main question being investigated and the research findings for the studies listed below. For each, what question were the researchers trying to answer and what was found when the research results were examined? You should understand the question being asked, know what the researchers found, and be able to answer both general and specific questions about the material.

Blood and Wolfe: research on conjugal decision making, and the resultant "resource hypothesis"
National Crime Survey, National Family Violence Survey, and similar research about violence

KEY THEORETICAL PERSPECTIVES

The purpose of any theory is to help us to see the world more clearly and explain *why* things are they are, thus leading us to greater understanding. You should be familiar with the following, being able to explain it briefly or at length, give examples, and answer pertinent questions.

French and Raven: six power bases
Blood and Wolfe: resource hypothesis
relative love and need theory

Quotations to assess: Read the following quotations and then think about them, assessing or evaluating them in terms of the text material in this chapter. Do not worry about being right or wrong. Just read the quotation and then think about in terms of the concepts, theories, ideas, and research results covered in this chapter of the textbook.

"The fundamental concept in social science is Power, in the same sense in which Energy is the fundamental concept in physics."
Bertrand Russell [Earl] (1872-1970), British philosopher, mathematician. *Power*. ch. 1, "The Impulse to Power" (1938).

"If Mr. Vincent Price were to be co-starred with Miss Bette Davis in a story by Mr. Edgar Allan Poe directed by Mr. Roger Corman, it could not fully express the pent-up violence and depravity of a single day in the life of the average family."
Quentin Crisp (b. 1908), British author. *Manners from Heaven*, ch. 2 (1984).

"However sugarcoated and ambiguous, every form of authoritarianism must start with a belief in some group's greater right to power, whether that right is justified by sex, race, class religion, or all four. However it may expand, the progression inevitably rests on unequal power and airtight roles within the family."
Gloria Steinem (b. 1934), U.S. feminist writer, editor. *Outrageous Acts and Everyday Rebellions*, "If Hitler Were Alive, Whose Side Would He Be on?" (1983; first published in Ms., New York, Oct/Nov. 1980).

"Power is not an institution, and not a structure; neither is it a certain strength we are endowed with; it is the name that one attributes to a complex strategical situation in a particular society."
Michel Foucault (1926-84), French philosopher. The History of Sexuality, vol. 1, pt. 4, ch. 2 (1976).

TRUE-FALSE

If you have studied and understand the textbook, you should be able to answer correctly the following true-false questions. Some of the questions are general and some are specific. Each is either obviously true or obviously false. None of the questions are tricky. Answer all of the questions by printing a T or F in the answer space for each question, and then check your answers with the correct answers at the end of this chapter. If you find that you miss questions, review the textbook material to discover *why* your answer is incorrect

_____ 1. Power that an individual exercises over others is "social power."

_____ 2. Power between married partners is referred to as conjugal power.

_____ 3. According to the resource hypothesis, the spouse with more resources has more power in the marriage.

_____ 4. Researchers have found that the higher the education and occupational status of husbands, the greater their conjugal power.

_____ 5. The resource hypothesis paid attention to which partner makes the most final decisions in a relationship.

_____ 6. American marriages tend to be inegalitarian, or, to exhibit institutionalized, unequal power differences.

_____ 7. In our society, we have an ideology of marital equality, but male dominance continues to prevail.

_____ 8. It may be a myth that patterns of power among blacks and Mexican-Americans are very different than those among non-Hispanic whites.

_____ 9. The principle of least interest refers to the tendency of school boards to elect as administrative disciplinarians those persons students find both boring, awe-inspiring and fear-inspiring.

_____ 10. The relative love and need theory assumes that women are as likely to have power as men are.

_____ 11. The relative love and need theory predicts that husbands will generally be more powerful.

_____ 12. Historically, women have tended to rely on "micromanipulation."

_____ 13. Lipen-Blumen argues that men dominate the public spheres of work and political leadership but that women tend to dominate the private sphere.

_____ 14. A study by Blumenstein and Schwartz, of married heterosexual couples, cohabiting couples, and lesbian couples, found that gender was by far the most significant determinant of the pattern of power in these relationships.

_____ 15. Marital violence exists in all social classes.

_____ 16. More programs/services exist to serve the needs of gays to cope with domestic violence than exist for other groups.

_____ 17. Marriage counselors are for the most part committed to helping couples avoid no-power relationships.

_____ 18. Wife-beating occurs more often in marriages where the wife can express herself better than can her husband.

_____ 19. No marriage is entirely free of power politics.

_____ 20. Battered wives' lack of personal power begins with fear, and that fear is frequently not unfounded.

_____ 21. Since Freddie likes Sarah Louise "a little bit" and Sarah Louise is deeply in love with Freddie, Freddie is in the position of having his way most or all of the time, regardless of Sarah Louise's desires to the contrary. If she doesn't "go along with him," he threatens to leave her. This illustrates the principle of least interest.

_____ 22. Loretta tells her alcoholic husband Sam that if only he won't drink, she will give him all of the sex he wants, any time he wants it, and however he wants it. Loretta is trying to use what the textbook calls "reward power" to change Sam's drinking.

MULTIPLE CHOICE

If you have studied and understand the textbook, you should be able to answer correctly the following multiple choice questions. Some of the questions are general and some are specific. Print the correct alternative letter-answer in the space to the left of each question. Check your answers with the correct answers at the end of this chapter. If you understand the chapter very well, you should miss few or none of these questions. For each question you miss, review the pertinent textbook material to understand *why* your answer is incorrect.

_____ 1. In the purest sense (as explored by the text), "power" is:
a. the ability to require people to do what you want them to do
b. something that is granted by the norms of religious or quasi-religious rituals
c. something that is granted by the legal norms
d. roughly equivalent to "influence"

_____ 2. Social power is the kind of power that an individual exercises:
a. among the upper-middle and upper classes
b. over oneself in public, in accordance with social norms
c. over oneself, and is similar to self-discipline or self-control, except it is social
d. over others

_____ 3. Research by Blood and Wolfe found that the more _________ one has, the more power one has over making decisions in families.
a. will power
b. flexibility
c. futurity
d. resources

_____ 4. In their research, Blood and Wolfe found that:
a. black husbands tended to have more conjugal power than did white husbands
b. younger spouses tended to make more decisions than did older spouses
c. slightly more families were wife-dominated than were husband-dominated
d. most families had a relatively equalitarian decision-making structure

_____ 5. The Blood and Wolfe research about family decision-making structure has been criticized for:
a. failing to consider the effect of social class on decision-making
b. ignoring the impact that resources have on who gets to make final decisions
c. not considering the extent to which education has an effect on family decision-making
d. assuming that the patriarchal power structure has been replaced by egalitarian marriages

_____ 6. The resource hypothesis encourages people to see conjugal power as:
a. a personality-based variable
b. a resource that is hereditary in the same sense that parental educational level is hereditary
c. equivalent to other kinds of power in its long-term, general impact on behavior
d. shared rather than patriarchal

_____ 7. According to the text, _______________ may be a myth.
a. black matriarchy
b. white patrimony
c. Hispanic avuncularity as expressed in La Familia
d. minority group endogamy

_____ 8. The marriages of blacks tend to be more ______ than those of whites.
a. free-form
b. egalitarian
c. matriarchal
d. innovative

_____ 9. According to the text, "micropolitics" is something that goes on:
a. within families
b. within one's own mind
c. within neighborhoods
d. at community-wide, state, and regional levels, and is primarily institutional

_____ 10. As the text uses the term, _________ seek to negotiate and compromise, not to "win."
a. segmental power couples
b. no-power couples
c. patriarchal couples
d. autocratic couples

_____ 11. Which of these means that both partners exert about equal power?
a. no-power
b. quasi-equilibrated power
c. centered power
d. scattered power

_____ 12. Among the victims of intimate violence, about 50 percent of the victims are attacked by:
a. husbands and/or wives
b. persons they have met only that same day as the attack
c. boyfriends and/or girlfriends
d. ex-spouses

_____ 13. In "neutralizing power," the subordinate:
   a. has counterarguments
   b. refuses to cooperate, takes a neutral position
   c. matches blow for blow and insult for insult with her/his opposition
   d. quote past happenings in the relationship, which most of the time results in the the opponent withdraw from arguing

_____ 14. Each year, about one out of every ____ couples in the United States, an individual commits at least one violent act against his or her partner.
   a. two
   b. three
   c. four
   d. six

_____ 15. During the entire length of a marriage, a violent act is committed in ____ percent of the couples.
   a. 35
   b. 48
   c. 60
   d. none of the above

_____ 16. According to the text, it is women aged ______ who have the highest rates of violent victimization attributable to intimates.
   a. 10-19
   b. 20-29
   c. 20-34
   d. 24-39

_____ 17. The most common form of family violence is violence between:
   a. mother and child
   b. mother and father
   c. father and child
   d. none of these

_____ 18. A husband forcing sex upon a wife, at least while they are living together, is legal in about what proportion of the states?
   a. about one-half
   b. about three-fifths
   c. about four-fifths
   d. nearly 85 percent

_____ 19. What is the third stage in the cycle of domestic violence referred to in the textbook?
   a. after an mild argument, the husband is tired and still angry
   b. the husband explodes into anger and physically abuses the wife
   c. the husband expresses regret and appears to be genuinely contrite
   d. the wife's reluctance to accept apologies angers the husband to be more violent

_____ 20. Johnson suggests that there are two forms of heterosexual violence against women. One of them is:
a. common couple violence
b. alienated violence
c. compensation violence
d. none of the above

_____ 21. According to research, which of the following types of violence is specifically mentioned in the textbook as being likely to lead to voluntary counseling?
a. violence among newly married couples
b. violence among couples married for a long time
c. violence in which now-adult children abuse a dependent elderly family
d. violence among lesbian couples

_____ 22. Which of the following is *not* a type of child neglect?
a. failing to provide guidance to child
b. failure to provide needed immunizations
c. being overly harsh and critical
d. none of the above

_____ 23. Shared living arrangements, caregiver's poor emotional health, and pathological relationship between victim and abuser are predictors of abuse rates for which of the following categories?
a. dependent husband and caregiving wife
b. caregiving husband and dependent wife
c. caregiving teen and dependent single parent
d. caregiving now-adult child and dependent elderly parent

Your Opinion, Please: *How much hitting is O.K.? How much verbal attack is O.K.?*
Please don't over-react to these questions. Consider them seriously.

Why ask the questions at all? **Because they are issues and won't go away** just because we prefer not to acknowledge them as issues and/or consider them resolved because we think we for sure know the answer to the questions.

Should parents be allowed to spank their children at all? How much of a spanking is "too much?" Some people feel that no physical punishment should be allowed. Some of our ancestors felt that parents had to right to put an unruly child to death. Today social workers and the courts have not yet resolved the question of the point at which physical discipline become child abuse and mistreatment.

How much should one spouse be allowed to hit another if at all? It serves no purpose to ignore the fact that some family research has been conducted which did not count as spouse abuse one spouse slapping the other spouse. Slugging with an open fist was counted as abuse, but slapping was not in this research considered to be unusual or unacceptable behavior—or, as spouse abuse.

The problem is compounded when cultures and societies other than our own are considered. You can find your own examples, but to get your started, recall that there are thousands of bride-burnings in India (lack of payment of the dowry, or failure to pay increased dowry resulting in "accidental" kitchen fires many of which prove fatal to the woman).

Of course, the essence of law is to bring order to society when there are those whose behavior is unacceptable to others, and who will not conform willingly, so the power of the courts is brought into play to extract conformity, punish offenders, and/or to offer rehabilitation .

What is your view about this issue? Do you take an absolute stand (this behavior is wrong and that's all there is to it)? Do you take a relativistic stand (the behavior is to be judged from the perspective of the norms of those engaging in the behavior)? Or do you support some other point of view? Give a reasoned argument supporting your perspective, whatever it may be.

SHORT-ANSWER ESSAY QUESTIONS

The following are sample short-answer essay questions — questions of the type you may be asked if your instructor uses questions like these. Even if your instructor does not use questions like these, you can help organize and consolidate your learning if you can answer these questions in a well-organized and complete manner.

Do not think that it is easier to write adequate answer for brief essays than for longer essays. Short-answer essays may be more challenging because answers are required to be both brief and complete. And after you have answered these questions below, construct some similar questions of your own devising, and then answer them. The Study Questions at the end of each textbook chapter make very good short-answer essay questions.

1. Distinguish between personal power, social power, and conjugal power. Give an example of each, making sure that it is clearly from your examples why they are good examples.

2. Briefly but completely define Blood and Wolfe's resource hypothesis and two criticisms of it.

3. Summarize the three-phase cycle of violence.

4. How does abuse and its consequences differ among lesbian and gay male couples?

5. How extensive is child abuse? Summarize the most important research results and findings on this issue.

6. What measures or programs have been implemented to combat child abuse? What are the points of disagreement among experts regarding these measures or programs?

7. What characterizes elder abuse and elder neglect?

ESSAY

The following are sample essay questions — questions of the type you may be asked if your instructor uses essay questions. Even if your instructor does not use essay questions, you can help organize and consolidate your learning if you can answer these questions in a well-organized and complete manner. Of these essay questions, the third essay question is usually the most challenging.

1. Construct a hypothetical couple that you make-up or construct such that they are likely candidates for be physically abused by or physically abuse the other spouse. Your constructed hypothetical example should embody as much as possible of what is known from research findings on this topic.

2. Write a well-reasoned essay in which you set forth seven reasons abused wives "put up with" wife abuse. If you list the reasons, be sure that for each you include enough information to make it clear that your are thoroughly familiar with the "reason why."

3. Explore in detail the parallels and differences between elder abuse and other forms family violence. Include in your answer appropriate research results to make your answer more persuasive.

ANSWERS

CHAPTER SUMMARY

Power, the ability
emotional **child abuse**
and **child abuse**
elder abuse and neglect

COMPLETION

1. power
2. no-power
3. principle of least interest
4. neutralizing power
5. personal power
6. resource hypothesis
7. conjugal power
8. no-power
9. neutralizing power
10. informational power

TRUE-FALSE

1. T
2. T
3. T
4. T
5. T
6. T
7. T
8. T
9. F
10. T
11. F
12. T
13. T
14. T
15. T
16. F
17. F
18. T
19. T
20. T
21. T
22. T

MULTIPLE CHOICE

1. A
2. D
3. D
4. D
5. D
6. D
7. A
8. B
9. A
10. B
11. A
12. C
13. B
14. D
15. D
16. C
17. D
18. A
19. C
20. A
21. D
22. D
23. D

CHAPTER 11

TO PARENT OR NOT TO PARENT

This chapter explores the causes and consequences of individual and collective decisions and behavior related to fertility rates—factors that play a part in both parenting and population size, growth and decline. The chapter moves from examination of differential fertility rates among the general population, African Americans, Hispanics, Asian Americans, and Native Americans, to exploration of factors affecting decisions to parent or not to parent. American couples, it seems, increasingly have at least three emerging options to add to the traditional list: remaining child-free, postponing parent, and the one-child family. Additional complications are added by issues and facts related to pregnancy, it occurrence, continuation through childbirth, and possible termination. Involuntary infertility has been influenced, as have other issues, by rapid changes in reproductive technology that have given rise to social and ethical issues. Adoption also enters into discussion and decisions about parenting. The chapter ends by exploring the options and responsibilities of decisions related to parenting, in all its complexities.

CHAPTER SUMMARY

Today, individuals have more choice than every about whether, when, and how many children to have. Although parenthood has become more of an option there is no evidence of an embracing childlessness. The majority of Americans continue to value parenthood, believe that childbearing should accompany marriage, and feel social pressure to have children. Only a very small percentage view childlessness as an advantage, regard the decision not to have children as a positive, believe that the ideal family is one without children, or expect to be childless by choice.

Nevertheless, it is likely that changing values concerning parenthood, the weakening of social norms prescribing marriage and parenthood, a wider range of alternatives for women, the desire to postpone marriage and childbearing, and the availability of modern contraceptives and legal abortion will eventually result in a higher proportion of Americans remaining childless. In fact, some observers have begun to worry that American society may be drifting into a period of structural antinatalism.

Children can add a fulfilling and highly rewarding experience to people's lives, but they also impose complications and stresses, both financial and emotional. Couples today are faced with options other than the traditional family of two or more children, remaining childless, postponing parenthood until they are ready, and having only one child. Often, people's decisions concerning having a family are made by default.

Birthrates are declining for married women. Especially white women are waiting longer to have their first child and are having, on the average, two children. Although pregnancy outside of marriage has increased, many unmarried pregnant women choose _________. There are many social and ethical issues surrounding the technologies and behaviors related to contraception and abortion.

Open and closed ________ are ways of becoming a parent without conceiving; some families have both adopted and biological children.

Point to Ponder: Every society faces this difficult issue: How to replace people who die and so have enough people to do what needs to be done to keep society going. Most societies achieve this by sexual reproduction or by immigration. But some societies have found they are under-reproducing —more people are dying or leaving than are being born or emigrating to the society, so the society's population is shrinking. Examples include the Island of Malta and the nation of Hungary. How long can a society afford to under-reproduce or shrink in population size? And what can be done to make it possible for society to survive under such conditions? Since fertility rates are sensitive to uncertainties and to economic conditions, does this suggest a role for government (tax breaks for having children, lower interest rates for home loans if applicants are parents, etc.)? Does the U.S. government have policies that serve this purpose?

KEY TERMS

You should be able to explain the concepts listed below. In your explanation, try to avoid using the concept you are explaining. You should be able to *give several examples* of each concept and to *explain why* each example is an example.

total fertility rate 308
total fertility rate 306
pronatalist bias 311
structural antinatalism 312
opportunity costs 314
programmatic postponers 316
paradoxical pregnancy 326
involuntary infertility 331
subfecundity 331
selective reduction 333
public adoption 338
private adoption 338
closed adoption 339
open adoption 339
disrupted adoptions 341
dissolved adoptions 341
attachment disorder 341

COMPLETION

Complete the following sentences by selecting the correct alternative from the Key Terms listed above. Some key terms may be used more than once. Some may not be used at all. Filling in a blank may require more than one word.

1. ________________________ are those adoptions that take place through licensed agencies that place children in adoptive families.

2. Disruption and dissolution rates sometimes are associated in children developing ________________________________, whereby they "shut off" willingness or ability to make future attachments to anyone.

3. The ________________________________ is the number of births a typical woman will have over her lifetime.

4. The ______________________________________ is a cultural perspective or attitude in which having children is taken for granted and not having children must be justified.

5. Some first-time parents are ____, who deliberately put off having their first child because of some mutual intention, consciously arrived at by the couple, whereby they know exactly why they have chosen to delay having their first child.

6. A ____________________________________ is one had by women who were guilty and disapproving about having premarital sex, and were simultaneously the least likely to use contraception regularly in association with their sexual behavior.

7. ________________________________ is the expulsion of the fetus or embryo from the uterus either naturally or medically (spontaneously or induced).

8. ____________________________________, or secondary infertility, refers to parents who have difficulty having additional children.

9. Another phrase for selective termination is ____, in which some but not all fetuses in a multiple pregnancy due to ovulation-stimulating drugs are selectively aborted.

10. In a(n) ________________________________, the process involves some direct contact between the biological and adoptive parents, ranging from one meeting before the child is born to lifelong friendship.

KEY RESEARCH STUDIES

You should be familiar with the main question being investigated and the research findings for the studies listed below. For each, what question were the researchers trying to answer and what was found when the research results were examined? You should understand the question being asked, know what the researchers found, and be able to answer both general and specific questions about the material.

U.S. Bureau of the Census: data related to fertility, including differences by ethnic/racial groups
U.S. Bureau of the Census: data related to abortion, timing of abortions, and related issues

Quotations to assess: Read the following quotations and then think about them, assessing or evaluating them in terms of the text material in this chapter. Do not worry about being right or wrong. Just read the quotation and then think about in terms of the concepts, theories, ideas, and research results covered in this chapter of the textbook.

"There are no illegitimate children, only illegitimate parents—if the term is to be used at all."
Bernadette Devlin McAliskey (b. 1947), Northern Irish politician. Quoted in: *Irish Times* (Dublin, 31 July 1971).

"Parents lend children their experience and a vicarious memory; children endow their parents with a vicarious immortality."
George Santayana (1863-1952), U.S. philosopher, poet. *The Life of Reason*, "Reason in Society," ch. 2 (1905-6; rev ed., 1953).

"The one regret I have about my own abortions is that they cost money that might otherwise have been spent on something more pleasurable, like taking the kids to movies and theme parks."
Barbara Ehrenreich (b. 1941), U.S. author, columnist. *The Worst Years of Our Lives*, "Their Dilemma and Mine" (1991; first published 1989).

"It is now quite lawful for a Catholic woman to avoid pregnancy by a resort to mathematics, though she is still forbidden to resort to physics and chemistry."
H.L. Mencken (1880), U.S. journalist. *Notebooks*, "Minority Report" (1956).

"The cemetery of the victims of human cruelty in our century is extended to include yet another cemetery, that of the unborn."
Pope John Paul II [Karol Wojtyla] (born. 1920), Polish ecclesiastic, pope. Quoted in: *Observer* (London, 9 June 1991).

"What is a neglected child? He is a child not planned for, not wanted. Neglect begins, therefore, before he is born."
Pearl S. Buck (1892-1973), U.S. author. *Children for Adoption*, ch. 3 (1964).

TRUE-FALSE

If you have studied and understand the textbook, you should be able to answer correctly the following true-false questions. Some of the questions are general and some are specific. Each is either obviously true or obviously false. None of the questions are tricky. Answer all of the questions by printing a T or F in the answer space for each question, and then check your answers with the correct answers at the end of this chapter. If you find that you miss questions, review the textbook material to discover *why* your answer is incorrect

_____ 1. The U.S. fertility rate is the number of births a typical woman will have over her lifetime.

_____ 2. U.S. fertility has dropped, and childbearing has increasingly shifted to later ages.

_____ 3. The white fertility rate is now approximately 1.97.

_____ 4. African Americans have currently have the highest fertility rate of any racial/ethnic minority in the U.S.

_____ 5. Native American births have fallen sharply since 1970, in contrast to the general racial/ethnic minority trend.

_____ 6. In 1992 in the U.S., about 16 percent of women age 30 were still childless.

_____ 7. A pronatalist bias refers to a bias in favor of contraception and abortions, and in opposition to forced unwanted pregnancy and childbirth.

_____ 8. Foregoing lost wages, pleasant vacation time, and career continuation or advancement in order to complete a pregnancy and have a child is referred to as the "sunk costs" of pregnancy and childbearing.

_____ 9. Slightly over one-half of U.S. women who were still childless but intended to marry and/or have a child said that they intended to have only one child during their childbearing years.

_____ 10. Only children are not more lonely, spoiled, selfish, and dependent that children who have siblings.

_____ 11. The double standard assigns responsibility for contraception to both sexes.

_____ 12. The percentage of children born out of wedlock, or, to unmarried parents, is currently higher than it has ever been since 1940.

_____ 13. In 1993, nearly 70 percent of African American births occurred outside marriage.

_____ 14. In 1993, births to teen mothers were about 26 percent of all births in the U.S.

_____ 15. One promising "control drug" for HIV/AIDS is a drug with the acronym STD.

_____ 16. *Roe v. Wade* legalized abortion throughout the United States, but it did not legalize any and all abortions.

_____ 17. According to recent and compelling evidence, abortion has a statistically significant ability to make further pregnancies problematic in many cases, difficult in most cases, and impossible in some cases.

_____ 18. Subfecundity is a term that refers to the parents with one child having difficulty becoming pregnant again.

_____ 19. Legally adopted children are currently about 11 percent of all children in any given year.

_____ 20. African Americans are now about 12 percent of the U.S. population, but about 20 percent of infants and children awaiting adoption are of African-American parentage.

MULTIPLE CHOICE

If you have studied and understand the textbook, you should be able to answer correctly the following multiple choice questions. Some of the questions are general and some are specific. Print the correct alternative letter-answer in the space to the left of each question. Check your answers with the correct answers at the end of this chapter. If you understand the chapter very well, you should miss few or none of these questions. For each question you miss, review the pertinent textbook material to understand *why* your answer is incorrect.

_____ 1. The U.S. total fertility rate reached the lowest level ever recorded in this country in:
   a. 1856
   b. the early 1930's, during the Great Depression
   c. 1976
   d. 1993, a response to national and international uncertainties

_____ 2. The tendency for people with high education and/or income is for them to have relatively fewer children, and this pattern is typical of:
   a. "Old American" (multiple generations in the U.S.) families only
   b. the white population, immigrant or not, but it is not typical for "people of color"
   c. the population for which English is spoken in the family home
   d. all racial groups in the U.S.

_____ 3. The racial/ethnic group with the highest fertility in the U.S. is:
   a. immigrant nonwestern European Catholics
   b. Asian Americans
   c. African Americans
   d. none of these

_____ 4. The Native American fertility rate in the U.S. today is:
   a. approximately what it was at the time of the Great Frontier
   b. lower than that of Asian Americans
   c. lowering significantly and persistently
   d. none of these

_____ 5. Some population experts explain the recent history of the U.S. fertility rate by taking the position that American society is:
   a. becoming aggressively pronatalist
   b. becoming antinatalist
   c. favorable to parenting but not to children, and many potential parents don't choose to bear children until society changes significantly to be more "child-friendly"
   d. none of these

_____ 6. Children present parents with high direct costs and:
   a. high opportunity costs
   b. lower opportunity costs
   c. opportunity costs that are low but rising as the fertility rate declines
   d. opportunity costs that are low but increasing as the fertility rate declines

_____ 7. Which of these is the opposite of being "pronatalist"?
   a. being for early marriage
   b. being against contraception
   c. supporting couples being child-free or childless
   d. affirming cohabitation and group marriage among the elderly

_____ 8. Child-free women tend to be:
   a. ambivalent toward children, but strongly bonded toward their spouse
   b. desirous of having children, but unable to find a marriageable man with whom to begin a traditional family
   c. highly traditional, but antinatalist
   d. attached to a satisfying career

_____ 9. Since the 1960's, birth control has been viewed mainly as:
   a. the woman's responsibility
   b. a shared responsibility that is negotiable, just as are other choices made knowledgeably
   c. the man's responsibility
   d. decisions made by default, except where prevention of disease is a consideration

_____ 10. The majority (_______percent in 1993) of all births outside marriage are to white mothers.
a. 13 percent
b. 35 percent
c. 42 percent
d. 60 percent

_____ 11. For black women, marriage and parenthood:
a. have become separate experience, not necessarily connected
b. remain inextricably connected as a norm, but disconnected as a fact
c. remain inextricably connected for the grandparent generation only
d. none of these

_____ 12. The increase in childbearing among older single women is largely a phenomenon among:
a. African Americans
b. Hispanics
c. Asian Americans
d. none of the above

_____ 13. It is ironic that the rates of premarital pregnancy are increasing:
a. at a time when teens are disavowing interest in family and/or parenthood
b. at a time when contraceptives are so available
c. at the time that drug use is increasing the percent of live births that are born "damaged"
d. at a time when abortion rates are undergoing a slight but continuing decline

_____ 14. Paradoxical pregnancies are pregnancies being undergone by persons who among their characteristics and attitudes include the fact that:
a. they come from families where such pregnancies have produced negative results
b. they tend to have negative, disapproving, guilt attitudes toward premarital sex
c. most of these pregnancies began by being wanted but ended by being unwanted
d. they have parental approval rather than disapproval

_____ 15. Research has found that for single women ages 14-21, ____________ was associated with increased likelihood that they will use birth control.
a. currently residing in a two-parent household
b. having sex with only one partner
c. living in a community with a high STD rate
d. being in school or employed

_____ 16. Ninety percent of all:
a. unmarried mothers reported using birth control at least half of the time
b. abortions occur within the first trimester
c. unmarried mothers go to marry the father of their child born out of wedlock
d. pregnancies are uneventful in that they result in uncomplicated childbirth

_____ 17. Abortion was legalized in:

a. 1973
b. 1985
c. 1991
d. none of the above

_____ 18. Selective reduction is a procedure by means of which:

a. the number of prospective mates is reduced over time in a semi-competitive process
b. the number of fertile days is determined by taking body temperature three times a day for a period of approximately one week
c. probability of having all female fetuses or all male fetuses is increased
d. none of the above

_____ 19. About ____ of U.S. adoptions are of children whose birth origin is outside the country. That is, they are "international" adoptions.

a. 5 percent
b. 15 percent
c. 25 percent
d. 35 percent

_____ 20. In semi-open adoptions, biological and adoptive families:

a. engage in exchange of personal information but do not have direct contact
b. have either both mothers or both fathers communicate with each other
c. exchange personal information and even see one another, but do not talk or become close acquaintances
d. none of these

Your Opinion, Please: Surrogate motherhood is not possible and in fact, occurs. But some ethical problems remain. For example, there is the race issue. Given the history of racial exploitation in this and other countries, is it appropriate for a woman of one race to carry and bear a child for a man of another race? Why does it matter? Or doesn't it? Going further, what if the mother and father of a fertilized egg are citizens of the Unites but the "pure" surrogate who carries the child during pregnancy is a citizen of another country? What should be the citizenship of the infant? What does the law say?

SHORT-ANSWER ESSAY QUESTIONS

The following are sample short-answer essay questions — questions of the type you may be asked if your instructor uses questions like these. Even if your instructor does not use questions like these, you can help organize and consolidate your learning if you can answer these questions in a well-organized and complete manner.

Do not think that it is easier to write adequate answer for brief essays than for longer essays. Short-answer essays may be more challenging because answers are required to be both brief and complete. And after you have answered these questions below, construct some similar questions of your own devising, and then answer them. The Study Questions at the end of each textbook chapter make very good short-answer essay questions.

1. Briefly but completely summarize fertility trends in the U.S. since 1800-1993, as set forth in the textbook.

2. Compare ethnic/minority fertility rates with the non-Hispanic white fertility rate. What similarities/differences can be seen?

3. Paulette and Alfred have been considering having only one child, their newborn daughter Crystie. Paulette thinks having one child is a dangerous situation for everyone, while Alfred thinks that having one child has benefits for everyone in the family. What does the textbook say that might better inform Paulette and Alfred about their dilemma and choices that may be made?

4. Are abortions safe? Be brief, specific, and complete.

5. What has been the recent history of RU-486 contraceptive, as detailed in the textbook?

6. What is selective reduction and for whom is it likeliest to emerge as an issue or decision?

7. How has the status of parenthood been affected by reproductive technology? (You should be able to write a complete answer using only the remainder of this page.)

ESSAY

The following are sample essay questions — questions of the type you may be asked if your instructor uses essay questions. Even if your instructor does not use essay questions, you can help organize and consolidate your learning if you can answer these questions in a well-organized and complete manner. Of these essay questions, the third essay question is usually the most challenging.

1. Compare each of the following three ethnic/racial minority groups' fertility rates with the fertility rate of non-Hispanic whites. Be systematic, detailed, and complete in your answer. At the end of your essay, *summarize* these differences in a well-worded paragraph.

2. Explore parents' costs and benefits of having children, making sure to include in your answer social pressures as one factor that should be considered.

3. Discuss the issues that characterize transracial and international adoption in U.S. society today.

4. Write a well-reasoned essay in which you discuss the three emerging options discussed by the textbook regarding parenthood: remaining childfree, postponing parenthood, and having one child. For each, be sure to discuss the pro's and con's inherent in selecting the option.

ANSWERS

CHAPTER SUMMARY

women choose ***adoption.***

open and closed ***adoption***

COMPLETION

1. public adoption
2. attachment disorder
3. total fertility rate
4. pronatalist bias
5. programmatic postponers
6. paradoxical pregnancy
7. abortion
8. subfecundity
9. selective reduction
10. open adoption

TRUE-FALSE

1. T
2. T
3. T
4. F
5. F
6. F
7. F
8. F
9. F
10. T
11. F
12. F
13. T
14. F
15. F
16. T
17. F
18. T
19. F
20. T

MULTIPLE-CHOICE

1. C
2. D
3. D
4. D
5. B
6. A
7. C
8. D
9. A
10. D
11. A
12. D
13. B
14. B
15. D
16. B
17. A
18. D
19. B
20. A

CHAPTER 12

PARENTS AND CHILDREN OVER THE LIFE COURSE

This chapter explores the general situation of parents and children over the life course of parents in modern America, amidst an amalgam of old and new images of parents and parenting. Parents are diverse and to be found in a variety of configurations, from traditional heterosexual partners to gay male parents.

Social class affects parenting in a changing economy, as parents who are in the poverty level, or who are blue-collar, lower middle-class, upper middle-class, or upper class parent in ways that are to some extent varied. The same may be said for racial/ethnic diversity and parenting. African Americans, Native Americans, Mexican Americans, Asian Americans raising minority children in a racist and discriminatory society all encounter somewhat different realities as they go about their parenting tasks.

The five stages of parenting and the five parenting styles are further testimony to the variety that is parenting in the United States these days, variety that both continues and changes in some respects when both parents and children grow older.

CHAPTER SUMMARY

Raising children is both exciting and frustrating. The ________________ theoretical perspective reminds us that society-wide conditions influence the relationship, and these factors can place extraordinary emotional and financial strains on parents. Formerly, children were expected to become more help and less trouble as they grew older; today, it is different. Most of what children need costs more as they grow: clothes, transportation, leisure activities, and schooling.

The chapter began by presenting some reasons why parenting and stepparenting can be difficult today. Some things noted are that work and parent roles often conflict. Middle-aged parents, especially mothers, may be a part of the ______________________________, caught been dependent children on one hand and increasingly dependent, aging parents on the other.

Not only mothers' but fathers' roles can be difficult, especially in a society like ours in which attitudes have changed so rapidly and in which there is no consensus about how to raise children and how mothers and fathers should parent. In addition, both children's and parents' needs change over the course of life. One thing that does not change as children mature is their need for supportive communication from a parent. Grandparents, particularly companionate and involved ones, can be helpful.

The need for supportive—and socially supported—parenting transcends social class and race or ethnicity. At the same time, we have seen that parenting differs in some important ways, according to economic resources, social class, and whether parent and child suffer discrimination due to minority status or sexual orientation of the parents.

To have better relationships with their children, parents need to recognize their own needs and to avoid feeling unnecessary guilt; to accept help from others (friends and the community at large as well as professional caregivers);and finally, to try to build and maintain flexible, intimate relationships using techniques suggested in this chapter and in the previous chapter.

> **Point to Ponder**: In some families, it is expected that children will work around the house and do chores without pay. It is felt in these families that a parent's role is to look out for a child and to provide spending money when a child seems to need it. Does the child provide unpaid labor that if done by an unrelated adult would be disallowed by law? If so, what justifies this? In the same way, what reasons justify hitting or withholding a meal from a child if the child's behavior does not meet with parental approval? Put differently, if society disapproves of these economic and disciplinary activities with adult recipients, how, specifically, does one account for the high approval rates for the use of these activities toward children in families?

KEY TERMS

You should be able to explain the concepts listed below. In your explanation, try to avoid using the concept you are explaining. You should be able to *give several examples* of each concept and to *explain why* each example is an example.

- sandwich generation 350
- shared parenting 350
- primary parents 350
- parenting alliance 354
- permissiveness 370
- overpermissiveness 370
- laissez-faire discipline 372
- autocratic discipline 373
- interactive perspective 373
- democratic discipline 374
- custodial grandparent 383
- noncustodial grandparent 383

COMPLETION

Complete the following sentences by selecting the correct alternative from the Key Terms listed above. Some key terms may be used more than once. Some may not be used at all. Filling in a blank may require more than one word.

1. Because of the status divorce and custody status of their own offspring, ____________________ may not get to see their own grandchildren very much.

2. A form of authority in which all family members involved have some say about what happens, and rules are discussed ahead of time, is termed ______________________________________.

3. In ________________________ discipline, parents let their children set their own goals, rules, and limits, with little or no parental guidance along those lines.

4.. A concern of parents in the ______________________________class is that their adult children do not have to more down in social class, and so leave their material and educational resources.

5. The co-parenting relationship of parents whereby they support each other (or sometimes unconsciously undermine each other) is termed a ______________________________.

6. ______________________________ is not determined in terms of time, but in terms of identity, in which a set of parent split the parenting responsibilities between themselves.

7. Middle-aged parents with both young adult children and with their own parents still alive are sometimes referred to as the ________________________________.

8. ____________________________________, an attitude of accepting the childishness of children, is fundamental to intimate parent-child relationships.

KEY RESEARCH STUDIES

You should be familiar with the main question being investigated and the research findings for the studies listed below. For each, what question were the researchers trying to answer and what was found when the research results were examined? You should understand the question being asked, know what the researchers found, and be able to answer both general and specific questions about the material.

U.S. Bureau of the Census and similar survey data: information related to fertility, including differences by ethnic/racial groups
LeMasters and DeFrain: research about blue-collar and lower middle-class parents
LeMaster and DeFrain: research about five parenting styles
Cherlin and Furstenberg: research about grandparents and grandparenting

Quotations to assess: Read the following quotations and then think about them, assessing or evaluating them in terms of the text material in this chapter. Do not worry about being right or wrong. Just read the quotation and then think about in terms of the concepts, theories, ideas, and research results covered in this chapter of the textbook.

"Parents ... are sometimes a bit of a disappointment to their children. They don't fulfill the promise of their early years."
Anthony Powell (b. 19050, British novelist. Stingham, in *A Buyer's Market*, ch. 2 (1952; second in the novel sequence, A Dance to the Music of Time).

"If you bungle raising your children, I don't think whatever else you do well matters very much." Jacqueline Kennedy Onassis (1929-1994) U.S. First Lady. Quoted in Theodore C. Sorenson, *Kennedy* Pt. 4, ch. 15 (1965).

"The most socially subversive institution of our time is the one-parent family."
Paul Johnson (b. 1928), British journalist. Quoted in: *Sunday Correspondent Magazine* (London, 24 Dec. 1989).

"Your responsibility as a parent is not as great as you might imagine. You need not supply the world with the next conqueror of disease or major motion-picture star. If your child simply grows up to be someone who does not use the world 'collectible' as a noun, you can consider yourself an unqualified success."
Fran Lebowitz, (b. 1951), U.S. journalist. Social Studies, "Parental Guidance" (1981).

"The thing that impresses me most about America is the way parents obey their children"
Edward, Duke of Windsor (1904-1972), King Edward VIII of Great Britain and Northern Ireland. Look (New York, 5 March 1957).

TRUE-FALSE

If you have studied and understand the textbook, you should be able to answer correctly the following true-false questions. Some of the questions are general and some are specific. Each is either obviously true or obviously false. None of the questions are tricky. Answer all of the questions by printing a T or F in the answer space for each question, and then check your answers with the correct answers at the end of this chapter. If you find that you miss questions, review the textbook material to discover *why* your answer is incorrect

_____ 1. The concept of childhood as different from adulthood did not emerge until about the seventeenth century.

_____ 2. In our society the parenting role (or stepparenting role) typically conflicts with the working role, and employers place work demands first.

_____ 3. Today's parents raise their children in a pluralistic society, characterized by relatively few views about ways to raise children.

_____ 4. The "sandwich" generation refers to those grandparents whose roles are restricted to basic but necessary custodial roles such as getting lunch ready for children to take to school.

_____ 5. Blue-collar parents are an emerging minority group.

_____ 6. "The parent as martyr" refers to a parent who has actually laid down his/her life for the child, actually willing to die instead of the child, as sometimes happens in relative rare medical settings or, less frequently, in complex accidents.

_____ 7. "Autocratic discipline" is associated with the police officer style of parenting.

_____ 8. "Blue-collar parents tend to empathize with their children's being creative, happy, and independent.

_____ 9. The children of upper-class parents are not necessarily guaranteed a comfortable level of living "no matter what."

_____ 10. Parents should intervene to restrain children from physical violence.

_____ 11. Parents should intervene to restrain children from physical violence.

_____ 12. Today's parents do not have clear guidelines bout what constitutes "good parenting."

_____ 13. Cultural pressure encourages adults to become parents even though they may not really want to do so.

_____ 14. Most mothers and fathers approach parenthood with little or no previous experience in child care.

_____ 15. Infants have different "readabilities." That is, they read at different grade levels as they develop--and some of them read at grade levels higher or lower than their chronological age.

_____ 16. The text seems to take the point of view that parental permissiveness toward children is a desirable thing.

_____ 17. Studies seem to show that violence between parent and child is the most pervasive form of family violence.

_____ 18. Surveys show that parents over age 65 prefer to live with their adult children so as to have their grandchildren nearby.

_____ 19. More black grandparents than white grandparents feel free to correct a grandchild's behavior.

_____ 20. The adult child involved in a parent's care heretofore was far more likely to be a daughter than a son--but research now reveals almost equal distribution of this kind of caregiving activity.

MULTIPLE CHOICE

If you have studied and understand the textbook, you should be able to answer correctly the following multiple choice questions. Some of the questions are general and some are specific. Print the correct alternative letter-answer in the space to the left of each question. Check your answers with the correct answers at the end of this chapter. If you understand the chapter very well, you should miss few or none of these questions. For each question you miss, review the pertinent textbook material to understand *why* your answer is incorrect.

_____ 1. Parents today are _____ to have to cope with serious childhood illnesses.
   a. much more likely than before
   b. much more likely than before
   c. as likely today as in previous years
   d. less likely

_____ 2. The average employed person is now on the job an additional ____ hours (a year than s/he was two decades ago) . . .
   a. 76
   b. 163
   c. 236
   d. 390

_____ 3. For a variety of reasons, parents today are probably judge by _____ standards than they were in the past.
   a. more uniform
   b. higher
   c. lower
   d. much lower

_____ 4. The "enduring image of motherhood" includes the idea that:
   a. a woman's identity is tenuous and trivial without motherhood
   b. the image of women is inextricably mingled with caregiving images from major religions
   c. significant others —including mothers— are indelible parts of that aspect of personality called the Generalized Other
   d. traditional images of mothers are perpetuated by the pervasive mass media such as television series and films

_____ 5. Studies indicate that when we control for class, the child-rearing style of African-American parents:
   a. compares to that of whites
   b. involves much harsher physical punishment when children misbehave
   c. allows children more unsupervised roaming through the neighborhood than white parents allow
   d. is more detached than the parenting style of white parents

_____ 6. E.E. LeMasters listed how many parenting styles?
   a. four
   b. five
   c. six
   d. seven

_____ 7. Which of these is NOT one of the parenting styles listed by LeMasters?
   a. the pal
   b. the police officer
   c. the teacher-counselor
   d. the servant

_____ 8. When LeMasters and DeFrain speak of "value stretch," they refer to which of these?
   a. compromising one's values
   b. valuing pushing oneself to the limit in terms of educational/occupational achievement
   c. getting the most out of every dollar spent in the household budget
   d. a difference in the parenting behaviors of blue-collar men and their wives

_____ 9. According to the text, the accomplishments of children of _____ parents, grandparents, and great-grandparents.
   a. upper-class
   b. upper middle-class
   c. lower middle-class
   d. working-class

_____ 10. According to Gelles and Straus, and Steinmetz, _____ can help reduce violence in families, especially between ______________.
   a. day care; grandparent and now-adult child
   b. day care; parent and preschooler
   c. well-trained police; husband and wife
   d. timeout; siblings

_____ 11. Cherlin and Furstenberg found that about _____ of their sample of grandparents said they saw their grandchildren less often than every two or three months.
   a. 77 percent
   b. 54 percent
   c. 31 percent
   d. 20 percent

_____ 12. As used in the text, the term "primary parents" refers to which of these?
   a. parents who are only beginning to learn basic parenting skills
   b. parents whose children are in primary school
   c. persons who are not the biological parents but who provide basic caregiving for one or more children
   d. couples in which both parents were equally involve din "mothering" behavior

_____ 13. A study of fathers in our society found that _____ thought they should share childrearing equally with mothers, but _____ *actually* did so.
   a. 55 percent; 68 percent
   b. 55 percent; 21 percent
   c. 74 percent; 13 percent
   d. 80 percent; 40 percent

_____ 14. According to the text, the concept "new fathers" is associated with which of these?
   a. first-time parents
   b. conversion-oriented parents
   c. 50-50 parents
   d. newly divorced fathers

_____ 15. Parents vary:
   a. in marital status but not in sexual orientation
   b. in marital status and in sexual orientation
   c. neither in marital status nor in sexual orientation
   d. none of the above

_____ 16. A study of adult children and elderly parents found that the adult children tended to let parents live independently as long as:
   a. they saw each other once in a while
   b. they saw each other at least four times a month, or, approximately weekly
   c. it was safe
   d. the parents were happy with the arrangements

_____ 17. According to the text, which of the following has had the effect of leaving many more grandmothers to assume the responsibility of child rearing?
   a. the spread of HIV/AIDS and use of crack cocaine
   b. legalization of spousal separation without specification of child support and/or child care
   c. greater use of marriage dissolution and less use of divorce
   d. increased commitment to education and career by both men and women

_____ 18. Parents who loan their children large amounts of money should consider the desirability of:
   a. giving goods, merchandise, or actual property rather than cash or negotiable instruments
   b. making the loan contingent on the promise of supportive care when the parent is old and frail
   c. formalizing such loans in writing and with the consultation of an attorney
   d. giving information about the loan to persons who are not family members, in order to avoid confusion in the event that the loan-making parent(s) die intestate (that is, having no legal will)

_____ 19. In our society, parenting ends when one's child reach age:
   a. 16-17
   b. 18-20
   c. 21-25
   d. none of these

_____ 20. In a long-term study of child development in Hawaii, the researchers spoke of a "self-righting tendency" of some children to:
   a. somehow have the prerequisites to emerge into adulthood in good shape
   b. choose political and social options that fit their personal circumstances, even if their decisions are marginally —yet still—legal
   c. make choices that result in minimal contact with representatives of formal authorities, whether school, work organizations, or law enforcement
   d. have an exaggerated idea of their own importance in the general scheme of things, alienating many people in the process

_____ 21. Which of these is a program stressing guidelines for no-win intimacy to the parent/child relationship?
a. LOVIT
b. PCLOVE
c. PET
d. AARP

_____ 22. A program offered in many communities to assist in effective parenting is:
a. STEP
b. MORE
c. NOVA
d. MARIGOLD

_____ 23. Looking for opportunities to show the child a new picture of him/herself, letting children overhear you say something positive about them, and being a storehouse for your child's special moments are all listed by the text as ways to:
a. make children assertive and effective
b. free children from playing roles
c. increase probability that children will comply with reasonable community expectations
d. enhance the self-image of children with diagnosed behavior problems

_____ 24. The _____ undercuts the child's development of his/her own competency; I contrast, the _____ does such things as assigns responsibility and sends positive messages about the child's competence and motivation.
a. peer group; parent
b. medical model; social-psychological model
c. ritualistic teacher; effective teacher
d. over-involved parent; consulting parent

_____ 25. According to the text, blacks are likelier than whites to:
a. use psychological punishment for children's misbehavior
b. use food as a reward for good behavior
c. become grandparents at an earlier age
d. rely on physicians rather than on books for suggestions about adequate parenting

_____ 26. "Overpermissiveness" allows undesirable ___ as ____.
a. feelings; motivations for actions
b. peers; role models
c. acts; expressions of feelings
d. role models; sources of socialization

_____ 27. According to the text, today's parents are more likely to be:
a. remaining in the same social class as their parents
b. experiencing social mobility, and thus coming into conflict with their own parents about the proper socialization of their children
c. caring for fragile infants
d. caring for parents who die of sudden, catastrophic illness

_____ 28. According to the text, today's well-educated and well-read parents:
a. have too many resources for their own children's well-being
b. are so fearful of possibilities that they often become too protective
c. may overreact to normal behavior
d. none of these

_____ 29. According to the text, parents within the same racial/ethnic group do not necessarily agree n the best approach to take to racial issues. About _____ of African-American parents do not attempt any explicit racial socialization.
a. one-eighth
b. one-fifth
c. one-half
d. one-third

_____ 30. "Zero-parent family" refers to a family in which:
a. a parent is physically present but is absent in any meaningful caregiving sense
b. a biological parent is not present in the home as a caregiver
c. both parents have established a "paper trail" in the criminal justice system, effectively nullifying their capability as an economically productive role model
d. the child remains with a biological parent who is not biologically the parent and who has not adopted the child in the usual way

Your Opinion, Please: Consider the following and ask yourself what relevance it has to parent/child relationships: The human species is "the only animal that laughs and weeps, ... the only animal that is struck by the difference between what things are and what they might have been."

SHORT-ANSWER ESSAY QUESTIONS

The following are sample short-answer essay questions — questions of the type you may be asked if your instructor uses questions like these. Even if your instructor does not use questions like these, you can help organize and consolidate your learning if you can answer these questions in a well-organized and complete manner.

Do not think that it is easier to write adequate answer for brief essays than for longer essays. Short-answer essays may be more challenging because answers are required to be both brief and complete. And after you have answered these questions below, construct some similar questions of your own devising, and then answer them. The Study Questions at the end of each textbook chapter make very good short-answer essay questions.

1. Is "parent" a social status" A social role? Both? Neither? What gives "parent" meaning? Biological parenthood? A social commitment? Or what? Be specific--not general--in your response.

2. Explain the meaning of "the parenting alliance."

3. How are never-married single mothers "different" from divorced, widowed, or separated mothers?

4. In what way(s) are some children "resilient"? Give supportive information and/or research results as part of your answer.

5. In what way(s) are gay male and/or lesbian parents different and in what does research reveal about the adequacy of their parenting.

ESSAYS

The following are sample essay questions — questions of the type you may be asked if your instructor uses essay questions. Even if your instructor does not use essay questions, you can help organize and consolidate your learning if you can answer these questions in a well-organized and complete manner. Of these essay questions, the third essay question is usually the most challenging.

1. Explain the ways in which parenting behavior differs or varies in the various social classes.

2. Of course, everyone is in many ways individual and unique. Nevertheless, couples and individuals can expect to go through often-observed stages as they raise their children. List and explain the stages set forth in the text.

3. How to the tasks of parents change as their children age? Be specific regarding age and with regard to the tasks mentioned or discussed.

4. How do the problems of parents with preschoolers differ from the problems of parents with young adult children? In your answer, go beyond simply saying that there are age differences. What other differences are there? If you can, cite specific research results to make your answer more persuasive.

5. Describe several features of our society, social structure, and culture that can make parenting especially difficult. (Note: Be sure that you do not get side-tracked into writing about individual *psychological* factors, though they are of course important factors when trying to answer other questions, possibly in other courses. This question is about how characteristics of society, culture, and social structure make parenting especially difficult.)

ANSWERS

CHAPTER SUMMARY

family ecology perspective remind us

the **sandwich generation**

COMPLETION

1. noncustodial grandparents
2. democratic discipline
3. laissez-faire
4. upper middle-class
5. parenting alliance
6. shared parenting
7. sandwich generation
8. permissiveness

TRUE-FALSE

1. T
2. T
3. F
4. F
5. F
6. F
7. T
8. F
9. F
10. F
11. T
12. T
13. T
14. T
15. F
16. T
17. F
18. F
19. T
20 F

MULTIPLE CHOICE

1. D
2. B
3. B
4. A
5. A
6. B
7. D
8. D
9. A
10. D
11. D
12. D
13. C
14. C
15. B
16. C
17. A
18. C
19. D
20. A
21. C
22. A
23. B
24. D
25. C
26. C
27. C
28. C
29. D
30. B

CHAPTER 13

WORK AND FAMILY

This chapter explores ways in which postindustrial society has through the invention of the labor force brought about dramatic changes in the traditional model of provider husbands and homemaking wives, producing both new options with a variety of rewards and costs. A large proportion of women in the labor force as paid workers has produced a wide variety of alternatives: two-career marriages, self-employment vs. part-time employment, shift work, as well as the phenomenon of persons leaving the labor force and then returning to it. The unexplored, unresolved issue of unpaid family work or housework has been brought abruptly to conscious consideration as a social issue. Families now struggle with a variety of ways to juggle employment and unpaid family work, amidst a variety of outcomes, leaving much to be resolved in the areas of work, family, family policy and social policy.

CHAPTER SUMMARY

The __________________ is a social invention. Traditionally, marriage has been different for men and women. The husband's job has been as breadwinner, the wife's as homemaker. These roles are changing as more and more women enter the work force. Women still remain very segregated occupationally and earn lower incomes than men, on the average.

The chapter distinguishes between two-earner and ____________________________ marriages. In the latter, wives and husbands both earn high wages and work for intrinsic rewards. Even in such marriages, the husband's career usually has priority. Responsibility for household work falls largely on wives. Many wives would prefer shared role, and negotiation and tension over this issue cast a shadow on many marriages. An incomplete transition to equality at work and at home affects family life profoundly.

The chapter emphasizes that both cultural expectations and public policy affect people's options. As individuals come to realize this, we can expect pressure on public officials to meet the needs of working families by providing supportive policies, parental leave, child care, and ____________.

Paid work is not usually structured to allow time for household responsibilities and women, rather than men, continue to adjust their time to accomplish both paid and unpaid work, such as "the second ________." "If we can't return to traditional marriage, an if we are not to despair of marriage altogether it becomes vitally important to understand marriage as a magnet for the strains of the stalled revolution" (Hochschild 1989,p. 18). To be successful, two-earner marriages will require social policy support and workplace flexibility. But there are some things the couple themselves can keep in mind that will facilitate their management of a working-couple family. Recognition of both positive and negative feelings and open communication between partners can help working couples cope with an imperfect social world.

Household work and child care are pressure points as women enter the labor force and the two-earner marriage becomes the norm. To make it work, either the structure of work must be changed, social policy must support working families, or women and men must change their household role patterns—and very probably all three.

Point to Ponder: Sometimes couples marry with the expectation that the husband will work full-time at a job he enjoys and that the wife will work part-time at a job she enjoys because she wants to bring some additional money into the household to make life a little more comfortable. But to their surprise, they find that to make ends meet, both may have to settle for full-time jobs that they do *not* particularly enjoy. Since this was not their agreement or expectation when they married, each partner may be deeply dissatisfied with the situation. Why might their economic situation come as a surprise to them? Perhaps more important, what should be done?

KEY TERMS

You should be able to explain the concepts listed below. In your explanation, try to avoid using the concept you are explaining. You should be able to *give several examples* of each concept and to *explain why* each example is an example.

labor force 391
good provider role 392
main/secondary provider couple 392
coprovider couple 392
ambivalent provider couple 392
househusbands 394
two-person single career 395
occupational segregation 397
pink collar jobs 398
two-earner marriages 400
two-career marriage
shift work 401
unpaid family work 402
second shift 405
stalled revolution 406
ideological perspective 407
rational investment perspective 407
resources hypothesis 407
reinforcing cycle 408
trailing spouse 415
commuter marriage 416
two-location families 416
child care 418
family day care 418
center care 418
elder care 420
family leave 420
flexible scheduling 420
job sharing 420
flextime 420
personal days 420
mommy track 421
life course solution 422
family-friend workplace policies 423
gender strategy 425
family myths 425

COMPLETION

Complete the following sentences by selecting the correct alternative from the Key Terms listed on the previous page. Some key terms may be used more than once. Some may not be used at all. Filling in a blank may require more than one word.

1. The notion that the husband or "man of the household" would provide all of the economic income was known as the ______________________________.

2. Men who are ______________________________ are those who stay at home to care for the house and family while his wife works.

3. Fred is expected to work 42 hours each week. But his office manager leaves it to him to decide how many hours and what days he will work to accumulate those 42 hours. This is the concept of ______________________________.

4. Intentionally created part-time jobs that are not particularly demanding are part of the ______________________________.

5. The ______________________________ refers to the time spent by two-earner couples in negotiating and fulfilling obligations at home, together, once employment time requirements have been met.

6. A(n) ______________________________ is one in which the spouses live and work in geographically distant places but do spend time with each other occasionally—perhaps on weekends or at other periodic intervals.

7. A(n)______________________________ means that a wife is meeting responsibilities providing for the household that are not clearly acknowledged.

8. In devising a(n) ______________________________, a person considers his/her goals and decides upon a course of action based on the expectations and norms of the culture to achieve those goals.

9. The concept of ______________________________ means that workers are allowed to take periods of time off from work—sometimes paid—to tend to family matters such as pregnancy, a parent's remarriage, terminal illness, or other major life event such as intensive child care.

10. A(n) ______________________________ is one who relocates to accommodate the other's (but not one's own) career.

11. The type of care called ______________________________ involves providing assistance with daily living activities to an elderly relative who is chronically frail, ill, or disabled. A large percent of employees miss work to provide this type of care.

12. When two employees share one position, this is known as ______________________________.

13. Many employees have the option of ____________________________ to take care of unexpected personal business. However, many employees are reluctant to exercise this option because they fear it may leave a slightly negative impression of their work reliability and attendance.

14. When a wife devotes herself to helping her husband do everything expected as socially appropriate as a homemaker, mother, wife, and also as a hostess for business associates, as a companion on company outings, and in other ways helping in ways that are expected or even essential for his occupational success, it is appropriate to speak of a ________________________________.

KEY RESEARCH STUDIES

You should be familiar with the main question being investigated and the research findings for the studies listed below. For each, what question were the researchers trying to answer and what was found when the research results were examined? You should understand the question being asked, know what the researchers found, and be able to answer both general and specific questions about the material.

U.S. Bureau of the Census: statistics on work and occupations, throughout the chapter

Hertz: research about two-career couples

Schwartz: research about the mommy track

KEY THEORETICAL PERSPECTIVES

The purpose of any theory is to help us to see the world more clearly and explain *why* things are they are, thus leading us to greater understanding. You should be familiar with the following, being able to explain it briefly or at length, give examples, and answer pertinent questions.

Karl Marx and others: theoretical perspectives on the origin, characteristics, structure and function of industrial and postindustrial society

Husbands, wives, employers, and types of compensation, from the perspectives of conflict theory and exchange theory

Quotations to assess: Read the following quotations and then think about them, assessing or evaluating them in terms of the text material in this chapter. Do not worry about being right or wrong. Just read the quotation and then think about in terms of the concepts, theories, ideas, and research results covered in this chapter of the textbook.

"This woman is headstrong, obstinate and dangerously self-opinionated."
Report by personnel Officer at I.C, I, rejecting Mrs. Margaret Thatcher for a job in 1948.

"No woman in my time will be Prime Minister or Chancellor or Foreign Secretary—not the top jobs. Anyway I wouldn't want to be Prime Minister. You have to give yourself 100%.
Margaret Thatcher (b. 1925), British Conservative politician, prime minister. Interview in *Sunday Telegraph* (London, 26 Oct. 1969), on her appointment as Shadow Education Spokesman.

"Roughly speaking, the President of the United States knows what his job is. Constitution and custom spell it out, for him as well as for us. His wife has no such luck. The First Lady has no rules; rather each new woman must make her own."
Shana Alexander (b. 1925), U.S. writer, editor. *The Feminine Eye*, "The Best First Lady" (1970; first published 1968).

"You have to know exactly what you want out of your career. If you want to be a star, you don't bother with other things."
Marilyn Horne (b. 1934), U.S. opera singer. Quoted in: Winthrop Sargeant, *Divas: Impressions of Six Opera Superstars* (1959).

"There are very few jobs that actually require a penis or vagina. All other jobs should be open to everybody."
Florynce R. Kennedy (b. 1916), U.S. lawyer, civil rights activist. Quote in: John Brady, "Freelancer with No Time to Write," in *Writer's Digest* (Cincinnati, Feb. 1974).

"Saving lives is not a top priority in the halls of power. Begin compassionate and concerned about human life can cause a man to lose his job. It can cause a woman not to get the job to begin with."
Myriam Miedzian, U.S. author. *Boys Will Be Boys*, ch. 2 (1991).

"Perpetual devotion to what a man call his business, is only to be sustained by perpetual neglect of many other things."
Robert Louis Stevenson (1850-94) Scottish novelist, essayist, poet. *Virginibus Puerisque*, "An Apology for Idlers" (1881).

TRUE-FALSE

If you have studied and understand the textbook, you should be able to answer correctly the following true-false questions. Some of the questions are general and some are specific. Each is either obviously true or obviously false. None of the questions are tricky. Answer all of the questions by printing a T or F in the answer space for each question, and then check your answers with the correct answers at the end of this chapter. If you find that you miss questions, review the textbook material to discover *why* your answer is incorrect

_____ 1. Today, occupational segregation and sex discrimination in employment persist, and the trend in this area is either a continuation or slight increase in the amount of such segregation and discrimination.

_____ 2. Working is inherently self-actualizing, but it must be meaningful to the worker, characterized by some appreciable degree of creative and self-direction.

_____ 3. American workers average hourly wages, when adjusted for inflation, have declined since 1973.

_____ 4. The U.S. Census no longer automatically considers or assumes the male to be the head of a household containing an adult male.

_____ 5. In an ambivalent provider couple, the husband embraces the main provider role, but does so begrudgingly, harboring feelings of resentment.

_____ 6. The good provider role incorporates both rewards and costs for men.

_____ 7. Society has encouraged men to give primacy to their work and to let their family relationships come second.

_____ 8. About 8 percent of husbands are currently househusbands.

_____ 9. Historically, the homemaker or housewife is a relatively modern role.

_____ 10. Mexican culture places more importance on the family than on work—for both sexes.

_____ 11. Full-time homemakers are permitted to set up IRAs (individual retirement accounts).

_____ 12. Research indicates that 55 percent of mothers of recently-born infants would prefer to seek meaningful outside employment than stay home with their infants.

_____ 13. The labor force participation of Latinas has traditionally been lower than that of white women.

_____ 14. Jobs tend to be significantly sex-segregated *between* occupational categories but not *within* occupational categories.

_____ 15. Two-earner marriages in which both partners are in the labor force are not yet the statistical norm.

_____ 16. A partner's doing shift work substantially increases the likelihood that fathers will participate in child care.

_____ 17. Employed wives do about twice as much housework as do their employed husbands.

_____ 18. About as few as one in five employed husbands do as much housework as their employed wives.

_____ 19. The pattern of spending considerably less time than women in housework is similar for all three racial/ethnic categories: non-Hispanic white, Hispanic, African American.

_____ 20. The second shift referred to in the textbook refers to those jobs that ordinarily begin at 3 p.m. and continue until 11 p.m., at which time the 3rd shift begins.

_____ 21. In the great majority of families, whoever *does* the household work *takes responsibility* for it, whether the task in question is done by the husband or by the wife.

_____ 22. Working mothers are said by the textbook's authors to be especially good role models for adolescent daughters.

_____ 23. Mothers who work part-time may actually spend more time helping their children with homework than do full-time homemakers.

_____ 24. A trailing spouse is one who attempts to use interstate information networks to attempt to locate a divorced spouse and so recover past-due child support.

_____ 25. Karen and Frederick are both employed on a full-time basis. Frederick keeps the garage tidy, mows the yard, and fixes electrical, plumbing, and carpentry tasks around the house when these problems arise. Karen, however, usually spends 15-25 hours per week doing things such a the household laundry, preparing meals and clean up after them, vacuuming, dusting the books on the bookshelf, the lamps, and also sees to it that the children have needed school materials. Karen puts in a "second shift," in the sense that term is used in the textbook.

MULTIPLE CHOICE

If you have studied and understand the textbook, you should be able to answer correctly the following multiple choice questions. Some of the questions are general and some are specific. Print the correct alternative letter-answer in the space to the left of each question. Check your answers with the correct answers at the end of this chapter. If you understand the chapter very well, you should miss few or none of these questions. For each question you miss, review the pertinent textbook material to understand *why* your answer is incorrect.

_____ 1. The trend is clearly away from distinguishing work based on:
- a. skill levels
- b. pay or wage levels
- c. sex or gender
- d. seniority

_____ 2. A serious "cost" of the good-provider role is/was that:
- a. the statuses attached to the role were confused and overlapping
- b. gender identification depended only on this one role
- c. gender identification became almost impossible, though sex identification was untouched
- d. the fines and court judgments were too expensive to continue with the role

_____ 3. When a wife spends much time and energy helping her husband in his career, this is what is meant by a(n):
- a. facilitating couple
- b. two-person single career
- c. sideline wife
- d. "organizer"

_____ 4. Society encourages men to give primacy to their ___ and let ___ come second.
- a. wife; children
- b. children; wife
- c. work; fictive kin
- d. work; family

_____ 5. The last women to move into employment outside the home have been:
- a. women recently widowed
- b. divorced women
- c. young women
- d. mothers of young children

_____ 6. The pronounced tendency for men and women to be employed in different types of jobs is called:
   a. job prejudice
   b. ability sifting
   c. mixed-ability job segregation
   d. occupational segregation

_____ 7. The concept associated with househusbands is:
   a. dysfunctional deviance
   b. latent avoidance
   c. role reversal
   d. gender exploitation

_____ 8. In a study of mothers of infants up to three months after their infants' births, researchers found that _____ of mothers felt that they would rather stay home with the baby than seek outside employment.
   a. 20-25 percent
   b. 40-45 percent
   c. 50-55 percent
   d. 70-75 percent

_____ 9. The "second shift" is connected in your text with:
   a. two-job husbands
   b. two-job wives
   c. the leisure gap
   d. beginning to have one's children later rather than earlier

_____ 10. About _____ of workers have flexible schedules, allowing them to choose their work hours—within some reasonable limits.
   a. 12 percent
   b. 22 percent
   c. 35 percent
   d. 48 percent

_____ 11. In the United States, historically, ________ have been more likely to work for wages than have __________.
   a. wives of upper white-collar men; wives of lower white-collar men
   b. wives of lower white-collar men; wives of upper white-collar men
   c. well-educated women; less well-educated women
   d. African-American women; white women

_____ 12. According to the text, women-dominated professions tend to be:
   a. professions with seasonal work
   b. occupations requiring lower rather than higher amounts of decision making
   c. service or support professions
   d. high-tech, "clean" occupations

_____ 13. The vast majority of single women are employed:
   a. as a temporary stage viewed as a prelude to marriage—in Japan, termed "diving-board girls"
   b. in order to supplement the economic requirements of their parental household
   c. in order to support themselves
   d. as a distraction to boredom while they wait for "the standard package" of marriage and family life

_____ 14. In the United States, _____ of husbands do as much housework as their wives do.
   a. 75-80 percent
   b. nearly 70 percent
   c. about half
   d. one-tenth or fewer

_____ 15. Hochschild uses the concept "stalled revolution" to refer to which of these?
   a. equalizing the gender difference in income for similar jobs
   b. employers providing child care at the work site
   c. amount of government concern about all employees' rights for domestic leave, regardless of sex
   d. increases in husbands' time spent doing household work

_____ 16. According to the text, a "reinforcing cycle" is associated with men's and women's:
   a. work
   b. involvement in child rearing
   c. boundary maintenance in gender roles
   d. responsibilities for children's formal educational experiences

_____ 17. According to the text, which of the following is sometimes or often an issue affecting a two-career family?
   a. needs for independence and autonomy
   b. geography
   c. medical needs
   d. none of these

_____ 18. The concept of a "trailing spouse" refers to a spouse who:
   a. relocates to meet the requirements of the other's career
   b. is the last to finish formal educational requirements for job advancement
   c. faces insecurities reducing her/him to using professional agencies to check up on the other's activities
   d. advances at the same rate occupationally but at a slower rate socially than the other spouse

_____ 19. As an alternative to the "mommy track," one analyst has suggested the "lifc course solution," which would require _____ the pattern of education, employment, and retirement.

a. reversing the current priorities of
b. destroying and rebuilding
c. rethinking and adjusting
d. none of the above

_____ 20. The adjustment of individual family members, as well as harmonious family relationships, requires _____ the very different and often conflicting needs of family members.

a. balance between
b. legitimization of
c. formal legal recognition of
d. elimination of

_____ 21. Paid child care costs between __________of the family budget.

a. 10-15 percent
b. 10-35 percent
c. 20-22 percent
d. 20-28 percent

Your Opinion, Please: Ted, an executive for an electronics company in Ohio, is married to Carol, who is not employed, as the couple agreed before they married. Ted's company decided to relocate his job to California. Carol, though not thrilled with the idea moved with him to California where she opened what turned out to be a highly successful mail-order business from their home. Ted's company then transferred his job back to Ohio. Carol saw no compelling reason to relocate, asking Ted to remain in California with her, since her income was now double his, and she refused to relocated to Ohio. Ted returned to Oho and sued his wife for divorce on grounds of desertion. The court awarded him a divorce on these grounds. What is your opinion about this case? Do you think Ted should have remained in California? Do you think Carol should have returned to Cleveland? Do you think Ted's court case was appropriate or do you see a more desirable alternative?

SHORT-ANSWER ESSAY QUESTIONS

The following are sample short-answer essay questions — questions of the type you may be asked if your instructor uses questions like these. Even if your instructor does not use questions like these, you can help organize and consolidate your learning if you can answer these questions in a well-organized and complete manner.

Do not think that it is easier to write adequate answer for brief essays than for longer essays. Short-answer essays may be more challenging because answers are required to be both brief and complete. And after you have answered these questions below, construct some similar questions of your own devising, and then answer them. The Study Questions at the end of each textbook chapter make very good short-answer essay questions.

1. For husbands, what are the rewards and costs of the provider role?

2. What are the differences in the reasons single women and married women are in the labor forces?

3. What is the “life course solution” to day care? Explain briefly but completely.

ESSAY

The following are sample essay questions — questions of the type you may be asked if your instructor uses essay questions. Even if your instructor does not use essay questions, you can help organize and consolidate your learning if you can answer these questions in a well-organized and complete manner. Of these essay questions, the third essay question is usually the most challenging.

1. Describe—and then compare and contrast—the work realities of husbands and of wives.

2. When couples change from being one-earner couples to being two-earner couples, what are the positive and negative consequences for them?

3. Cecelia and Andre are both fully employed professionals who intend to marry. They both intend to continue their professional careers. What do sociologists who study families know that would probably be useful for Cecelia and Andre? Be specific, be complete, and cite specific research findings to increase your answer's credibility.

CHAPTER SUMMARY

the ***labor force***
and ***two-career*** marriages
and ***flextime***

such as the second ***shift***
the option of ***personal time***
speak of a ***two-person single career***

COMPLETION

1. good-provider
2. househusbands
3. flextime
4. mommy track
5. second shift
6. commuter marriage
7. ambivalent provider couple
8. gender strategy
9. family leave
10. trailing spouse
11. elder care
12. two-person single career

TRUE-FALSE

1. T
2. T
3. T
4. T
5. F
6. T
7. T
8. F
9. T
10. T
11. T
12. F
13. T
14. F
15. F
16. T
17. T
18. F
19. T
20. F
21. F
22. T
23. T
24. F
25. T

MULTIPLE CHOICE

1. C
2. B
3. B
4. D
5. D
6. D
7. C
8. D
9. C
10. A
11. D
12. C
13 C
14. D
15. D
16. A
17. B
18. A
19. C
20. A
21. B

CHAPTER 14

MANAGING FAMILY STRESS AND CRISES

This chapter begins by discussing crisis, stressor, and stressor overload. Family transitions can be stressors: the sandwich generation, the postparental period, retirement, as well as widowhood and widowerhood. Family crises have expected, typical, or usual courses, beginning with a period of disorganization and often culminating in a period of recovery. Of course, families differ in the ways and effectiveness of coping with crisis as the adapt or try to adapt to the crisis by appraising the situation, and tap crisis-meeting resources. Meeting crises creatively involves as number of factors that have proved to be important determinants of the outcome of crises: having a positive outlook, spiritual values and support groups, high self-esteem, open and supportive communication, adaptability, having informal social support, having an helpful extended family, and the use of community resources.

CHAPTER SUMMARY

Family crises may be expected and normative, as when a baby is born or adopted, or they may be unexpected. In either case the event that causes the crisis is called a _________________________ . Stressors are of various types and have varied characteristics. Generally, stressors that are expected (normative), brief, external, defined by concrete norms, and improving are less difficult to cope with.

The predictable changes of individuals and families—parenthood, mid-life transitions, postparenthood, _________________________, and widow- and widowerhood—are all family transitions that can be viewed a stressors. During transitions, spouses can expect their relationship to follow the course of a family crisis.

A common pattern can be traced in families reactions to crises. Three distinct phases can be identified: the event that causes the crisis, the period of _________________________ that follows, and the reorganizing or recovery phase after the family reaches a low point. The eventual level of reorganization a family reaches depends on a number of factors, including the type of stressor, the degree of stress it imposes, whether it is accompanied by other stressors, and the family's definition of the crisis situation. Various models of family crisis and reorganization try to capture this process in an analytical way, to pursue through research useful knowledge about coping with crisis.

Meeting crises creatively means resuming daily functioning at or above the level that existed before the crisis. Several factors can help: a positive outlook, _____________________ values, the presence of support groups, high _________________________, open and supportive communication within the family, adaptability, counseling, and the presence of a kin ____________.

Point to Ponder: Think for a moment about the activities singles take part in as they date, go out, or get together. About what percent of these activities, which eventually may lead to marriage, give any clue as to how well or how poorly a potential spouse handles crises? In most cases, crisis-handling ability is something that does largely unobserved until after marriage or cohabitation has occurred. What effects do you think this has on sustaining such relationships? Can you think of a way to ensure that crisis-handling ability can be observed before marriage occurred?

KEY TERMS

You should be able to explain the concepts listed below. In your explanation, try to avoid using the concept you are explaining. You should be able to *give several examples* of each concept and to *explain why* each example is an example.

family stress 436
crisis 436
transitions 436
predictable 436
crises 436
stressor 437
stressor overload 438
sandwich generation 442
empty nest 443
bereavement 446
period of disorganization 448
ABC-X model 449
double ABC-X model 449
pile-up 449
resiliency model of family stress 452
adjustment 452
adaptation 452
strong families 453
weak families 453

COMPLETION

Complete the following sentences by selecting the correct alternative from the Key Terms listed above. Some key terms may be used more than once. Some may not be used at all. Filling in a blank may require more than one word.

1. A ______________________________ necessarily involves change.

2. Carol and Joe have a pleasant life with man pleasures and few strains. One day Joe's boss told him that further promotions will require that Joe get a Master's degree in the field of Business Administration. Joe's boss's announcement is an example of a(n):__________________.

3. Not only did Joe enter school but his mother died, which interrupted her support of Joe's education. One of the Carol and Joe's children needed to spend the winter in Spain for an academic exchange program. Carol lost several of her major clients at the advertising agency, through no fault of her own, and eight other similar things made life difficult for this couple. They are suffering from ____.

4. Carol was expecting that Joe would be required to get more education. Joe shared this expectation but hoped that it would be later rather than sooner. The fact that they expected it helped define it as a(n) ________________________________.

5. Characteristically, ____ creeps up on people without their realizing it.

6. The ________________________________ is much like the double ABC-X model, and elaborates on it.

7. "Family pile-up" (prior family hardships and strains that continue to affect family life) is part of the ____________________________________.

8. The ____ states that the stressor event (or) interacts with the family's ability to cope with a crisis (or B), which interacts with the family's definition of the event, which produces the crisis.

9. In coping with a crisis, it is in the ____ that family members face the decision of whether to express or to smother any angry feelings they may have.

10. The ____ is one transition with a high likelihood of a positive outcome.

11. The death of a much-loved grand-parent can ordinarily be expected to be followed by ________________________________.

12. Physical, mental, emotional, and intellectual symptoms, perception of declining health and depressive symptoms, along with anger, guilt, and sadness, are all associated with ______________________________.

KEY RESEARCH STUDIES

You should be familiar with the main question being investigated and the research findings for the studies listed below. For each, what question were the researchers trying to answer and what was found when the research results were examined? You should understand the question being asked, know what the researchers found, and be able to answer both general and specific questions about the material.

Research results about various specific types of crisis, disorganization, coping, and recovery, as discussed throughout the chapter:

- the sandwich generation
- the postparental period
- retirement
- widowhood and widowerhood
- alcoholism
- communicating with the dying

KEY THEORETICAL PERSPECTIVES

The purpose of any theory is to help us to see the world more clearly and explain *why* things are they are, thus leading us to greater understanding. You should be familiar with the following, being able to explain it briefly or at length, give examples, and answer pertinent questions.

Hill: the ABC-X model
McCubbin and Patterson: the double ABC-X model
H. and M. McCubbin: the resiliency model of family stress, adjustment, and adaptation

Quotations to assess: Read the following quotations and then think about them, assessing or evaluating them in terms of the text material in this chapter. Do not worry about being right or wrong. Just read the quotation and then think about in terms of the concepts, theories, ideas, and research results covered in this chapter of the textbook.

Crisis. An attributive noun. Often used to modify another noun, as in "crisis intervention," "family crisis," or "crisis center."
Origin: from Middle English, from Latin, from Greek word *krinein*, meaning "to separate."

"Written in Chinese the word crisis is composed of two characters. One represents danger and the other represents opportunity."
John F. Kennedy (1917-63), U.S. president, Speech, 12 April 1959, Indianapolis, Indiana.

"There can't be a crisis next week. My schedule is already full."
Henry Kissinger (b. 1923), U.S. Republican politician, Secretary of State. *New York Times Magazine* (1 June 1969).

"There is no lonelier man in death, except the suicide, than that man who has lived many years with a good wife and then outlived her. If two people love each other there can be no happy end."
Ernest Hemingway (1899-1961), U.S. author, *Death in the Afternoon*, ch. 11 (1932).

"Weeping may endure for a night, but joy cometh in the morning."
Hebrew Bible, Psalms 30:5.

"In this world without quiet corners, there can be no easy escapes from history, from hullabaloo, from terrible, unquiet fuss."
Salman Rushdie (b 1947), Indian-born British author. Outside the Whale (1984).

"As for types like my own, obscurely motivated by the conviction that our existence was worthless if we didn't make a turning point of it, we were assigned to the humanities, to poetry, philosophy, painting—the nursery games of humankind, which had to be left behind when the age of science began. The humanities would be called upon to choose a wallpaper of the crypt."
Saul Bellow (b. 1915), U.S. novelist, The Adventures of Augie Marsh, ch. 6 (1949).

TRUE-FALSE

If you have studied and understand the textbook, you should be able to answer correctly the following true-false questions. Some of the questions are general and some are specific. Each is either obviously true or obviously false. None of the questions are tricky. Answer all of the questions by printing a T or F in the answer space for each question, and then check your answers with the correct answers at the end of this chapter. If you find that you miss questions, review the textbook material to discover *why* your answer is incorrect

_____ 1. A crisis is a time of relative instability.

_____ 2. Common events can precipitate a crisis.

_____ 3. The ABC-X is applicable to an *individual*, whereas the double ABC-X model is applicable to a married *couple* or to a cohabiting *pair.*

_____ 4. The loss of potential children through miscarriage has all of the elements of the text's definition of a crisis.

_____ 5. The third phase of a crisis s the period of disorganization.

_____ 6. Alcoholism as a family crisis presents alcoholism as a crisis with seven stages.

_____ 7. Families who define a problem as their fault suffer less as individuals and also tend to provide less support than families who consider the cause to be external, probably because defining it as their fault provides at least some feeling of control over the situation.

_____ 8. Current research paints a picture of mothers in the empty-nest stage as being predominantly depressed and as having free-floating anxiety, a reaction that is not unusual but that is almost unvarying negative in its implications for mental health.

_____ 9. In a study of seventy-four husbands with multiple sclerosis, most of them responded to their disease by becoming "spectators" in their own homes.

_____ 10. As more women work outside the home, couple's adjustments to retirement may go somewhat more smoothly.

_____ 11. The better a couple adjusts to retirement, the less painful may be a forthcoming transition to widow- or widowerhood.

_____ 12. Widowhood is significantly less common in our society than is widowerhood.

_____ 13. Bereavement is a period of mourning.

_____ 14. Strong relationships with friends are related to high morale among widows.

MULTIPLE CHOICE

If you have studied and understand the textbook, you should be able to answer correctly the following multiple choice questions. Some of the questions are general and some are specific. Print the correct alternative letter-answer in the space to the left of each question. Check your answers with the correct answers at the end of this chapter. If you understand the chapter very well, you should miss few or none of these questions. For each question you miss, review the pertinent textbook material to understand *why* your answer is incorrect.

_____ 1. A crisis is defined by the fact that:
   a. it has a distinctly negative outcome
   b. it has the potential for a distinctly negative only
   c. it is a dramatic change from usual family relationships
   d. it is not caused by the family or by its individual members

_____ 2. In a crisis situation, there is:
   a. the prospect that things will never again be the same
   b. a period of relative instability
   c. the probability that family relationships will change for the worse
   d. a perception by the general community that the family is in some ways deviant

_____ 3. According to the text, a family stressor:
   a. occurs when the family is not ready to meet it because of overload
   b. has the potential to precipitate a crisis
   c. occurs when the family has to distinguish between two categories of events
   d. happens because outside events are always affecting the family

_____ 4. A "family pile-up" is part of:
   a. most people's life, especially during early marriage
   b. the familial culture lag model of family development
   c. the double ABC-X model of family crisis
   d. the pre-stressor syndrome

_____ 5. External stressors are:
   a. those that the family creates for itself
   b. those that are external to the individual but not the family itself
   c. those that come to the family from the external environment
   d. those that come from the physical environment rather than from the psychological environment

_____ 6. The concept of "pile-up" refers to which of the following?
   a. experiencing multiple stressors
   b. having several children, not just two
   c. having many bills to pay and paying them all at once
   d. family reunions

_____ 7. Families who _____ suffer more as individuals and also tend to provide less support for one another.

a. oppose instrumental relationships
b. have never faced a real crisis before
c. define a problem as their fault
d. have lower levels of formal education

_____ 8. In alcoholism as a family crisis, the third stage is:

a. attempt to eliminate the problem
b. denial of the problem
c. disorganization
d. reorganization with sobriety

_____ 9. In alcoholism as a family crisis, the fifth stage is:

a. attempts to eliminate the problem
b. denial of the problem
c. disorganization
d. efforts to escape the problem

_____ 10. A family crisis:

a. usually follows and A,B,C,A,B,C,A,B,C pattern
b. a fairly predictable pattern or course
c. ordinarily comes at either daily intervals, or monthly intervals, but not weekly ones
d. is usually unique for each family, but there are strong similarities *within* racial/ethnic groups

_____ 11. The word "crisis" comes from the Greek word for:

a. inflexibility
b. fright, or fear, or that which creates extreme anxiety
c. lack of practical experience
d. decision, choice, or separation into possible alternatives

_____ 12. According to the text, ________ whites are likelier to live in extended family groupings than are whites in the other social classes.

a. lower-class
b. pink-collar class
c. working-class
d. white-collar class

_____ 13. According to the text, _____ has a _____.

a. helping; dark side
b. crisis management; downward spiral
c. crisis management; mind-set based on an industrial model
d. ethnicity; negative effect on coping ability

_____ 14. According to the text, _____ more than ______ have difficulty coping with their severely retarded children.

a. older parents; younger parents
b. black parents; white parents
c. remarried couples' first marriage couples
d. fathers; mothers

_____ 15. Which of the following can help families after a crisis?

a. counseling
b. locating the person whose fault the crisis was
c. counseling the person whose fault the crisis was
d. making available a low-cost method to terminate family relationships

_____ 16. The resilience model of family adjustment and adaptation incorporates the "double ABC-X mod of family stressors and strains, but adds the elements of family system and family typology. Family typology is of two types:

a. strong or weak
b. out-reaching or inner-directed
c. resourceful or without coping skills
d. troubled or untroubled

_____ 17. According to the text, the "typical American family" has:

a. access to local governmental resources to which it fails to request access
b. access to state governmental resources to which it fails to request access
c. a low level of coping ability in areas of politics and economics
d. a high level of stress at all times

_____ 18. According to the text, even when it is deserved, ______ is less productive than viewing the crisis primarily as a challenge.

a. giving up
b. thinking innovatively
c. casting blame
d. seeking legal remedies

_____ 19. According to the textbook, __________ can serve as effective tools to cope with family stress and crisis.

a. the presence of healthy family pets
b. consumer behavior in stores and shops
c. rituals
d. none of these

_____ 20. According to the textbook, black married couples are no more likely than married couples to:

a. have extended crises
b. live with extended kin
c. solve violent crises effectively
d. solve emotional crises effectively

_____ 21. Women who adapt to husbands' unemployment by seeking emotional support from relatives and friends found that their husbands:

a. felt uncomfortable with the informal nature of the arrangements
b. insisted on formal rather than informal arrangements
c. were comforted by this indication of micro-community support
d. viewed this as disloyalty

_____ 22. A study of alcoholic families found that _____ can serve as effective tools to cope with family stress and crisis.

a. subcultural community support groups
b. pan-cultural value systems
c. rituals
d. informal contacts with community services agencies

_____ 23. Extended families consist of parents and their children who live in the same households with other relatives, such as the parents or a brother or sister of one of the spouses. Just under _____ percent of American family households are extended-family households.

a. 4
b. 9
c. 17
d. 23

_____ 24. ________________ may operate primarily as extended families, with shifting combinations of kin sharing economic resources, child-rearing responsibilities, and so forth.

a. Lower-class black communities
b. The families of clergy of various denominations
c. Studies cites in the text show that migrant laborers
d. none of the above

_____ 25. While in most racial/ethnic groups the extended family is viewed as a resource in times of trouble, some clinicians argue that in __________________ families, having the family find out about one's distress may be the problem rather than the solution.

a. working-class
b. upper-class
c. value-bonded
d. Irish

_____ 26. These days, increasingly, caregivers themselves are:

a. co-dependents
b. increasingly elderly
c. financially more able to support those dependent upon them
d. equal partners with the government in supporting those dependent on them

_____ 27. The text specifically mentions _____ as disrupting or weakening some family ties that might have been resources.

a. divorce
b. military relocation
c. compliance with government regulations
d. poor communication patterns

Your Opinion, Please: The notion that family problems ought to remain *family* problems — sometimes called privatization—may be a mixed blessing. It may ensure family privacy, but this privatization may make coping with crises more difficult. Which would you rather have: more family privacy but less help from friends and neighbors in coping with crises, or less family privacy but more help from friends and neighbors in coping with crises? Do you think it is possible to have both? If you think you can have both, how—precisely—is it possible to have both?

SHORT-ANSWER ESSAY QUESTIONS

The following are sample short-answer essay questions — questions of the type you may be asked if your instructor uses questions like these. Even if your instructor does not use questions like these, you can help organize and consolidate your learning if you can answer these questions in a well-organized and complete manner.

Do not think that it is easier to write adequate answer for brief essays than for longer essays. Short-answer essays may be more challenging because answers are required to be both brief and complete. And after you have answered these questions below, construct some similar questions of your own devising, and then answer them. The Study Questions at the end of each textbook chapter make very good short-answer essay questions.

1. What is a stressor and what is stressor overload? Give two detailed examples.

2. Explain briefly but completely how the postparental period can be a stressor.

3. Distinguish between "high self-esteem" and "having a positive outlook" as factors in meeting crises effectively.

4. Briefly but completely distinguish between strong families and weak familes, as discussed by McCubbin and McCubbin.

ESSAY

The following are sample essay questions — questions of the type you may be asked if your instructor uses essay questions. Even if your instructor does not use essay questions, you can help organize and consolidate your learning if you can answer these questions in a well-organized and complete manner. Of these essay questions, the third essay question is usually the most challenging.

1. Probably everybody has problems. But what is the distinction between having problems and have a crisis? Explore the distinction, making specific reference to material in the text.

2. What are the various stages of coming to grips with a crisis? And, do you think that people go through the same stages regardless of the crisis with which they are attempting to dope? Explore the question, but remember to support your answer with material from the text, because this is not an opinion question only..

3. Compare and contrast the ABC-X model, the double ABC-X model, and the resiliency model of family stress, adjustment, and adaptation.

CHAPTER SUMMARY

crisis is called a ***stressor***
postparenthood, ***retirement***
the period of ***disorganization***
spiritual values
high ***self-esteem***
of a kin ***network***

COMPLETION

1. crisis
2. stressor
3. stressor overload
4. transition
5. stressor overload
6. resiliency model of family stress, adjustment, and adaptation
7. double ABC-X model
8. ABC-X model
9. period of disorganization
10. empty nest
11. bereavement
12. bereavement

TRUE-FALSE

1. T
2. T
3. F
4. T
5. F
6. T
7. F
8. F
9. T
10. T
11. F
12. F
13. T
14. T

MULTIPLE CHOICE

1. C
2. B
3. B
4. C
5. C
6. A
7. C
8. C
9. D
10. B
11. D
12. C
13. A
14. D
15. A
16. A
17. D
18. C
19. C
20. B
21. D
22. C
23. A
24. A
25. D
26. B
27. A

CHAPTER 15

DIVORCE

This chapter begins by exploring how divorce rates are reported and examining the country's current divorce rate. The textbook reviewed why more couples are divorcing: decreased economic interdependence (the relationship between divorce and income; wives in the labor force), decreased social, legal, and moral constraints, high expectations for marriage, the changed nature of marriage itself, as well as other factors associated with divorce and some common marital complaints. Alternatives to divorce were reviewed. There was enumeration of the various stages of divorce: the emotional divorce, the legal divorce, the community divorce, and the psychic divorce. The economic consequences of divorce were found to include differences for husbands and for wives, various attitudes about alimony, and child support. The chapter reviews the effects of divorce on children, as well as "her divorce" and "his divorce," and ends by exploring consequences of divorce for the next generation.

CHAPTER SUMMARY

Divorce rates have risen sharply in this century, and divorce rates in the United States are now the highest in the world. In the past decade, however, they have begun to level off.

Reasons why more people are divorcing than in the past have to do with changes in society: economic ______ and legal, moral, and social_____ are lessening, expectations for intimacy are increasing, and expectations for _____ are declining. People's personal decisions to divorce, or to redivorce, involve weighing marital complaints—most often problems with communication or the emotional quality of the relationship—against the possible consequences of divorce. Two consequences that receive a great deal of consideration are how the divorce will affect children, if there are any, and whether it will cause serious financial difficulties.

The divorce experience is almost always far more painful than people expect. Bohannan has identified six ways in which divorce affects people. The six stations of divorce are the emotional divorce, the legal divorce, the ______ divorce, the psychic divorce, the economic divorce, and the _______ divorce. The psychic divorce involves a healing process that the individuals must complete before they can fully enter new intimate relationships.

The _______ divorce is typically more disastrous for women than for men, and this is especially so for custodial mothers. Over the past fifteen years, child support policies have undergone sweeping changes which are only now beginning to result in evaluation research.

Researchers have proposed five possible theories to explain negative effects of divorce on children. These include the life stress perspective, the parental loss perspective, the parental _____ perspective, the economic hardship perspective, and the interparental _____ perspective, this last of which asserts that conflict between parents before, during, and after divorce is responsible for children's lower well-being, is most strongly supported by research.

Husbands' and wives' divorce experiences, like husbands' and wives' marriages, are different. Both the overload that characterizes the wife's divorce and the loneliness that often accompanies the husband's divorce, especially when there are children, can be lessened in the future by more androgynous settlement. Divorce counseling can help make the experience less painful. Joint custody offers the opportunity of greater involvement by both parents, although its impact is still being evaluated. So also is the effect of parents' divorce on children's marital prospects.

Point to Ponder: Some welcomed "dissolution of marriage" statutes, which allowed spouses to petition the court for a dissolution of the marriage, as long as the husband and the wife agreed on disposition of assets, custody, and support. But research indicates that some spouses—probably wives more often than husbands—accept dissolution agreements that are not in their long-term best interests. If a spouse didn't realize at the time that the agreements were not in their best interests, do you think they should later have to live up to those agreements, or should they (perhaps by written agreement) left open for renegotiation at a later date?

KEY TERMS

You should be able to explain the concepts listed below. In your explanation, try to avoid using the concept you are explaining. You should be able to *give several examples* of each concept and to *explain why* each example is an example.

refined divorce rate 467
redivorce 468
structured separation 476
readiness for divorce 476
emotional divorce 476
divorce counseling 476
legal divorce 477
divorce mediation 477
no-fault divorce 477
relatives of divorce 479
psychic divorce 480
economic divorce 483
alimony 484
spousal support 484
rehabilitative alimony 484
displaced homemakers 484
entitlement 485
child support 485
Child Support Amendments (1984) 485
Family Support Act (1988) 485
guaranteed child support 486
children's allowance 486
life stress perspective 491
parental loss perspective 492
parental adjustment perspective 492
economic hardship perspective 492
interparental conflict perspective 494
custody 496
child-snatching 499
joint custody 499

COMPLETION

Complete the following sentences by selecting the correct alternative from the Key Terms listed on the previous page. Some key terms may be used more than once. Some may not be used at all. Filling in a blank may require more than one word.

1. ____________________ is kidnapping one's child from the other parent.

2. With _________________, used in France and Sweden, the government sends t the custodial parent the full amount of support awarded to the child, even though this sum may not have been received from the noncustodial parent. It is then the government's responsibility to collect the money from the parent that owes it.

3. The _____________________________ is the number of divorces per 1000 women over age 15.

4. Intensely hostile couples who say they think they want a divorce may try a period of __________ in which they live apart for a limited period, avoid getting lawyers, starting new relationships, and continue getting counseling.

5. In _________________, the ex-husband pays his ex-wife "just enough cash to get her back on her feet," but in many cases, this is not enough.

6. In _________________, one gains psychological autonomy through emotional separation from the personality and influence of the former spouse, learning to feel whole and complete again, and to have faith in one's ability to cope with the world.

KEY RESEARCH STUDIES

You should be familiar with the main question being investigated and the research findings for the studies listed below. For each, what question were the researchers trying to answer and what was found when the research results were examined? You should understand the question being asked, know what the researchers found, and be able to answer both general and specific questions about the material.

Research results about divorce as reported in U.S. Census data, the National Center for Health Statistics, and similar sources, as reported throughout the chapter.
Wallerstein and Kelly: research about children's postdivorce adjustment
Amato and Keith: research about children and young adult's postdivorce adjustments

KEY THEORETICAL PERSPECTIVES

The purpose of any theory or theoretical perspective is to help us to see the world more clearly and explain *why* things are they are, thus leading us to greater understanding. You should be familiar with the following, being able to explain it briefly or at length, give examples, and answer pertinent questions.

Paul Bohannan: six stations of divorce
Amato: five theoretical perspectives of effects of divorce on children
- life stress perspective
- parental loss perspective
- parental adjustment perspective
- economic hardship perspective
- interparental conflict perspective

Quotations to assess: Read the following quotations and then think about them, assessing or evaluating them in terms of the text material in this chapter.

"You can't stay married in a situation where you are afraid to go to sleep in case your wife might cut your throat."
Mike Tyson (b. 1966), U.S. boxer. Quoted in: *Daily Telegraph* (London, 1 Feb. 1989).

"Many divorces are not really the result of irreparable injury but involve, instead, a desire on the part of the man or woman to shatter the setup, start out from scratch alone, and make life work for them all over again. They want the risk of disaster, want to touch bottom, see where bottom is, and, coming up, to breathe the air with relief and relish again."
Edward Hoagland (b. 1932), U.S. novelist, essayist. "Other Lives," in *Harper's* (New York, July 1973); repr. *in Heart's Desire*, (1988).

"Being divorced is like being hit by a Mack truck. If you live through it, you start looking very carefully to the right and to the left."
Jean Kerr (b. 1923), U.S. author, playwright. Mary, in *Mary, Mary*, act 1.

"When two people decide to get a divorce, it isn't a sign that they 'don't understand' one another, but a sign that they have, at last, begun to."
Helen Rowland (1875-1950), U.S. journalist. *A Guide to Men*, "Divorces" (1922).

"The possibility of divorce renders both marriage partners stricter in their observance of the duties they owe to each other. Divorces help to improve morals and to increase the population."
Dennis Diderot (1713-1784), French philosopher. In *Selected Writings,* ed. by Lester G. Crocker, 1966).

TRUE-FALSE

If you have studied and understand the textbook, you should be able to answer correctly the following true-false questions. Some of the questions are general and some are specific. Each is either obviously true or obviously false. None of the questions are tricky. Answer all of the questions by printing a T or F in the answer space for each question, and then check your answers with the correct answers at the end of this chapter. If you find that you miss questions, review the textbook material to discover *why* your answer is incorrect

_____ 1. The number of divorces per year takes into account the general increase in the population.

_____ 2. The ratio of current marriages to current divorces is a faulty measure.

_____ 3. The crude divorce rate is the number of divorces per 1,000 population.

_____ 4. The most useful and valid divorce rate appears to be the refined divorce rate.

_____ 5. The higher the income, the less likely couples are to divorce.

_____ 6. Intriguingly enough, recent evidence indicates that divorce can be an uplifting, fulfilling experience—bordering on the romantic.

_____ 7. For all societies undergoing the transition from traditional to modern, divorce rates have increased.

_____ 8. Cohabitation is associated with higher probability of divorce.

_____ 9. Having parents who are divorced increases the likelihood of divorcing.

_____ 10. Adult children's relationships with their own parents generally do not change after the children's divorce.

_____ 11. Don't do your mourning for your divorce while you are going through your divorce; put it off until after the divorce has been concluded.

_____ 12. "Alimony" comes from the Latin word that means "to slowly diminish."

_____ 13. Parental divorce or some factor connected with it, is associated with lowered well-being of the adult children of divorce.

_____ 14. The "parental loss perspective" assumes that a family with both parents living in the same household is the optimal environment for children's development.

_____ 15. The divorce experience is almost always far more painful than people expect.

MULTIPLE CHOICE

If you have studied and understand the textbook, you should be able to answer correctly the following multiple choice questions. Some of the questions are general and some are specific. Print the correct alternative letter-answer in the space to the left of each question. Check your answers with the correct answers at the end of this chapter. If you understand the chapter very well, you should miss few or none of these questions. For each question you miss, review the pertinent textbook material to understand *why* your answer is incorrect.

_____ 1. In the 20th century the frequency of divorce has had dips and upswings associated with:
a. voter turn-out in presidential election years
b. historical events such as the Great Depression and World Wars
c. availability of commodities and foodstuffs, as determined by the market economy
d. the fecundity rate

_____ 2. The number of divorces per 1,000 population is the:
a. refined divorce rate
b. crude divorce rate
c. demographic divorce rate
d. none of the above

_____ 3. In this country's recent past, the divorce rate peaked or reached its highest point ever in:
a. 1955
b. 1967
c. 1979
d. 1993

_____ 4. Regarding household headed by men, this figure:
a. is highest for non-Hispanic whites
b. is lowest for non-Hispanic whites
c. is highest for African-Americans
d. varies little on the basis of race.

_____ 5. More likely to get a divorce are those who:
a. fail to complete college
b. have neutral to emotionally cool relationships with their own parents
c. have relatively lower than average verbal skills
d. commute longer rather than shorter distances to work

_____ 6. According to research results reported in the textbook, which of these increases the likelihood of divorce?
   a. both husband and wife having grandparents who divorced
   b. being of middle-class or lower middle-class Hispanic ethnic background
   c. have a post-high school education in a highly technical or "scientific" area
   d. husband and wife working in the same relatively small (20-50 employee) office environment

_____ 7. Low or deficient emotional qualities of marriage:
   a. tend to lead to divorce
   b. tend to lead to divorce mainly among those to who expect emotional satisfaction from marriage
   c. seem unrelated to the divorce rate
   d. are unrelated to the divorce rate unless the reason(s) for low emotional satisfaction are also directly related to some form of contact with the police or system of incarceration

_____ 8. What Bohannan called "emotional divorce" was associated with:
   a. levels of psychological disturbance or anxiety prior to the divorce
   b. behaviors exhibited in the courtroom when proceedings were underway
   c. divorce between spouses who had strong feelings *of whatever kind* toward each other
   d. none of the above

_____ 9. Almost all states have some statutes or laws guaranteeing:
   a. children the right to privacy in their own room, regardless of which custodial parent
   b. grandparents' visitation rights to see their grandchildren, under some circumstances
   c. children the right to refuse to exercise their access to gay male/lesbian non-custodial parents
   d. that gay male parents be allowed to be the custodial parent, positive or negative characteristics of the ex-wife notwithstanding

_____ 10. "Psychic divorce" refers to:
   a. an informal divorce, such as manifests itself in the instance of common law marriage
   b. divorces that take place "by proxy" in the way that some marriages formerly occurred by proxy
   c. attempted divorce by going to a medium or other person claiming special psychic powers, and achieving a mutual emotional/intellectual separation from the spouse in this manner
   d. none of these

_____ 11. Which of the following is one of the stations of divorce as recognized by Bohannan?

a. the mutual divorce
b. the children's divorce
c. the kinship system's divorce
d. none of these

_____ 12. To be successful a(n) _________________ requires a long period of mourning.

a. mortality divorce
b. divorce that has been a long time in coming
c. economic divorce
d. none of the above

_____ 13. A _________ study of divorce would have to study a cohort of married couples over a lifetime, collecting and analyzing data for the entire period.

a. longitudinal
b. randomized
c. stratified
d. none of the above

_____ 14. The high divorce rate:

a. does not mean Americans have given up on marriages; they want rewarding ones
b. now stands poised at just under 45 percent of all marriages
c. is probably an overestimate since it includes dissolutions of marriage in divorce statistics
d. reflects the fact that most divorces are in the middle years of marriage

_____ 15. "No-fault" divorce legally abolishes:

a. the need for expensive legal proceedings
b. legal proceedings at all; the couple simply go their separate ways
c. the concept of the "guilty party"
d. marriage ties, but does not abolish economic ties

_____ 16. "Displaced homemakers" are divorced women who:

a. literally find themselves out on the street, looking for new residence
b. find that someone else has taken their place in their ex-husband's affections
c. are older, full-time homemakers who suddenly find themselves divorced an without adequate support
d. are not given custody of their children; that custody goes to the husband and/or to the husband's new partner

_____ 17. The "sleeper effect" is found among:

a. girls coming from divorced families
b. divorced wives
c. teenage boys coming from divorced families
d. the timing of expenses related to getting divorced, about two-thirds of which expenses arrive more than six months after the divorce is legally granted

Your Opinion, Please: Imagine two people whose relationship culminated in marriage. They loved each other, felt the need for a shared future, and seemed perfectly suited for each other —and married. But people change. And not even marriage stops change. Often people who were suited to each other change in the same direction and remain well-matched. But sometimes people change in different directions so that one or both no longer want the same things and no longer feel comfortable with the idea of a shared future. If one partner changes in a direction that is difficult or unacceptable to the other partner, should they face the fact and seek a divorce? Or should they work at the relationship? Should one partner suppress his/her "new direction" for the sake of the current partnership and its shared past, or should they divorce rather than compromise in any way? Try to be specific in explain why you reason as you do in your answer to these questions. Can you think of some concrete examples from people you may happen to know?

SHORT-ANSWER ESSAY QUESTIONS

The following are sample short-answer essay questions — questions of the type you may be asked if your instructor uses questions like these. Even if your instructor does not use questions like these, you can help organize and consolidate your learning if you can answer these questions in a well-organized and complete manner.

Do not think that it is easier to write adequate answer for brief essays than for longer essays. Short-answer essays may be more challenging because answers are required to be both brief and complete. And after you have answered these questions below, construct some similar questions of your own devising, and then answer them. The Study Questions at the end of each textbook chapter make very good short-answer essay questions.

1. Distinguish between *crude divorce rate* and *refined divorce rate.*

2. Summarize the increases/decreases in the divorce rate since about 1920. Do not be overly general in your answer.

3. Briefly but completely summarize what research results tell us about *redivorce*.

4. What is *divorce mediation*?

5. Who are the *relatives of divorce?* Explain and given two examples.

6. Distinguish between *economic divorce* and *child support.*

7. Distinguish between *entitlement* and *guaranteed child support.*

8. Compare and contrast the *parental loss perspective* and *the parental adjustment perspective* regarding explanations of negative outcomes among children of divorced parents.

ESSAY

The following are sample essay questions — questions of the type you may be asked if your instructor uses essay questions. Even if your instructor does not use essay questions, you can help organize and consolidate your learning if you can answer these questions in a well-organized and complete manner. Of these essay questions, the third essay question is usually the most challenging.

1. More couples are divorcing or in other ways terminating their marriages. In a well-organized, thoughtful, thorough essay, explain why this is happening.

2. Material in the text suggests that many divorces go through various stages. Thought it may seem unique to the persons involved in "their own divorce," sociologists notice some of the same stages when comparing many divorces. Illustrate and explain in a well-organized essay, being both general and specific.

3. Louise and Andy considered divorcing for some time, and eventually got divorced. Write an essay in which you explore how people come to such a decision and what usually happens to them as they divorce. Explore briefly any ways in which Louise's divorce might be "different" from Andy's divorce.

CHAPTER SUMMARY

economic ***interdependence***
social ***constraints***
expectations for ***permanence***
the ***community*** divorce
and the ***coparental***
economic divorce is
the parental ***adjustment*** perspective
interparental ***conflict*** perspective

COMPLETION

1. childnapping
2. guaranteed child support
3. refined divorce rate
4. structured separation
5. rehabilitative alimony
6. psychic divorce

TRUE-FALSE

1. F
2. T
3. T
4. T
5. T
6. F
7. T
8. T
9. T
10. F
11. F
12. F
13. T
14. T
15. T

MULTIPLE CHOICE

1. B
2. B
3. C
4. D
5. A
6. C
7. A
8. D
9. B
10. D
11. D
12. D
13. A
14. A
15. C
16. C
17. A

CHAPTER 16

REMARRIAGES

This chapter begins by exploring some basic facts about remarriage, paying special attention to its effects of children's living arrangements. The "traditional exchange" in remarriage is reviewed regarding remarriage as it affects and is affected by children and the age of persons who remarry, along with homogamy in remarriage. Research results about the happiness, stability and well-being of remarries leads into a discussion of remarried families characteristics, kin networks, and family law as they affect remarried families. Moving next to stepparenting as a challenge in remarriage, the chapter discusses some reasons stepparenting id difficult, including financial strains, role ambiguity, stepchildren's hostility, as well as the stepmother and stepfather traps.

CHAPTER SUMMARY

Although remarriages have always been fairly common in the United States, patterns have changed. Remarriages are fare more frequent now than they were earlier in this century, and tend to occur after divorce more often than after ________. The courtship process by which people choose remarriage partners has similarities to courtship preceding first marriages, but the basic exchange often weighs more heavily against older women, and ______________ tends to be less important.

Second marriages are usually about as happy as first marriages, but they tend to be slightly less stable. An important reason is the lack of a cultural ________. Relationships in immediate remarried families and with kin are often complex, yet there are virtually no social perceptions and few legal definitions to clarify roles and relationships.

The lack of cultural guidelines is clearest in the ___________ role. Stepparents are often troubled by financial strains, role ______________, and stepchildren's hostility. Marital happiness and stability in remarried families are greater when the couple have strong social support, high expressiveness, a positive attitude about the remarriage, low role ambiguity, and little belief in negative stereotypes and myths about remarriages or stepfamilies. Personal remarriage agreements can help to establish understanding where few social norms exist.

Point to Ponder: According to the text, remarriages have a higher divorce rate than first marriages. Several possible reasons for remarriages' higher divorce rates are:

1. the social class of people who divorce and remarry
2. persons who remarry may be accepting of divorce—they have done it before
3. remarriages have stresses in addition to the usual stresses of marriage, including the presence of stepchildren

Can you think of other reasons? Give it some thought and write the reasons that occur to you.

KEY TERMS

You should be able to explain the concepts listed below. In your explanation, try to avoid using the concept you are explaining. You should be able to *give several examples* of each concept and to *explain why* each example is an example.

remarriages 510
double remarriages 520
single remarriages 520
remarried families 522
quasi-kin 523
coparenting 523
stepmother trap 529
hidden agenda 532

COMPLETION

Complete the following sentences by selecting the correct alternative from the Key Terms listed above. Some key terms may be used more than once. Some may not be used at all. Filling in a blank may require more than one word.

1. Conflicting expectations for behavior and attitude are the essential ingredients for the ________________________________.

2. The____________________________________ consists of partners in a remarriage failing to fully reveal their expectations for the partner in the remarried relationship.

3. Bohannan suggests the term ________________________________ to refer to the person who is married by one's former spouse. The prefix of the word actually means "somewhat" or "similar to" when used along with a suffix.

4. In ____________________________________, both partners have been married before.

5. Members of ________________________________ have no cultural script.

6. In a ____, one partner is marrying for the first time and the other partner is remarrying.

7. In a ________________________________, both partners are remarrying.

KEY RESEARCH STUDIES

You should be familiar with the main question being investigated and the research findings for the studies listed below. For each, what question were the researchers trying to answer and what was found when the research results were examined? You should understand the question being asked, know what the researchers found, and be able to answer both general and specific questions about the material.

Research results about remarriage as reported in U.S. Census and similar sources, as reported throughout the chapter.
Spanier and Furstenberg: well-being among the remarried
White and Booth: marital quality in first marriages and remarriages
Raphael, et al: the stepmother trip
Bohannan; Cherlin: general problems of stepparenting

Quotations to assess: Read the following quotations and then think about them, assessing or evaluating them in terms of the text material in this chapter.

step- prefix
Related by means of a remarriage rather than by blood: stepparent.
[Middle English, from Old English ***steop-***.]

digamy (dig′- a - me) noun
Remarriage after the death or divorce of one's first husband or wife. Also called **deuterogamy**.
— **digamous** adjective

The American Heritage Dictionary of the English Language, Third Edition, 1992, Houghton Mifflin Company and InfoSoft International, Inc.

TRUE-FALSE

If you have studied and understand the textbook, you should be able to answer correctly the following true-false questions. Some of the questions are general and some are specific. Each is either obviously true or obviously false. None of the questions are tricky. Answer all of the questions by printing a T or F in the answer space for each question, and then check your answers with the correct answers at the end of this chapter. If you find that you miss questions, review the textbook material to discover *why* your answer is incorrect

_____ 1. Remarriage is an alternative that more Americans are choosing.

_____ 2. The U.S. Supreme Court has ruled that failure to pay child support is not by itself enough to prevent a remarriage.

_____ 3. The average divorced person who remarries does so within six months to one year after the divorce.

_____ 4. Remarriages differ from first marriages in important ways.

_____ 5. In general, marriage favors husbands more than wives.

_____ 6. The "double standard of aging" works *for* women rather than *against* them in the *re*marriage market.

_____ 7. A factor that works against women in the remarriage market is the presence of children.

_____ 8. The divorce rate has leveled off, but the remarriage rate has declined steadily since 1966.

_____ 9. Generally, people who remarry after divorce are less willing to choose divorce as a way of resolving an unsatisfactory marriage.

_____ 10. Researchers have found that the presence of stepchildren is not a significant factor in the instability of marriages.

_____ 11. Quasi-kin are kin whose deviant behavior has resulted in distancing them from other family members.

_____ 12. The role of stepparent is less clearly defined than the role of parent in our society.

_____ 13. Stepchildren tend to be well-adjusted but do not get along with their stepfather as well as other children do with their natural fathers.

_____ 14. The so-called "hidden agenda" is a contradiction in terms, since it occurs when the participants have been *too* clear and *too* candid about their expectations for the remarriage relationship to work well.

MULTIPLE CHOICE

If you have studied and understand the textbook, you should be able to answer correctly the following multiple choice questions. Some of the questions are general and some are specific. Print the correct alternative letter-answer in the space to the left of each question. Check your answers with the correct answers at the end of this chapter. If you understand the chapter very well, you should miss few or none of these questions. For each question you miss, review the pertinent textbook material to understand *why* your answer is incorrect.

_____ 1. According to the text, the remarriage rate peaked most recently during which of the following periods?
   a. at the end of World War II
   b. at the end of the Great Depression
   c. in 1989
   d. in 1966

_____ 2. Which of the following is TRUE regarding the remarriage rate?
   a. The remarriage rate has declined steadily since 1966.
   b. Today, remarriages make up a relatively insignificant proportion of all marriages.
   c. The remarriage rate fell sharply during World War II.
   d. About 1/4 of the population who divorce eventually remarry.

_____ 3. About ______ of the people who divorce each year remarry eventually.
   a. one-fifth
   b. one-third
   c. two-thirds
   d. four-fifths

_____ 4. According to the textbook, fewer and fewer _________ see divorce and remarriage as a moral issue.
   a. African-Americans
   b. non-Hispanic whites
   c. Hispanics
   d. none of the above

_____ 5. Most likely to remarry, and having the highest remarriage rates, are:
   a. divorced men
   b. divorced women
   c. single men
   d. widowed women

_____ 6. __________ more than ____________ want to remarry.

a. Whites; Hispanics
b. Hispanics; Asian Americans
c. The young; the middle aged
d. Men; women

_____ 7. _______who remarried in the 1980s had higher annual family incomes than _______.

a. Black women; white women
b. Northeasterners; southwesterners
c. Church members; non- church members
d. none of the above

_____ 8. Marital happiness and ____________ are not the same, as has been pointed out more than once in the textbook.

a. sexual satisfaction
b. joy in being together as a couple
c. conjugal happiness
d. stability

_____ 9. Researchers consistently find ______ in marital happiness between first and later marriages.

a. little difference
b. a moderate positive relationship
c. a moderate negative relationship
d. none of the above

_____ 10. Our society offers members of remarried familes __________.

a. no rehabilitative alimony in case of divorce
b. no possibility of "dissolution of marriage," requiring instead adversarial divorce
c. no sexual scenario
d. no cultural script

_____ 11. Which of the following has a built-in difficulty or "trap"?

a. step-grandparent
b. the visiting biological parent
c. the stepfather
d. the stepmother

Your Opinion, Please: Some people remarry and re-divorce several times. The divorce rate is as high as it is because it reflects multiple remarriages and divorces. People who divorce are likelier to divorce again if they remarry. Why, in your opinion, does this happen? Be specific about the reason(s) you give. Do you think these reasons are the same reasons as for the first marriages and divorces, or do you think the reasons are different?

SHORT-ANSWER ESSAY QUESTIONS

The following are sample short-answer essay questions — questions of the type you may be asked if your instructor uses questions like these. Even if your instructor does not use questions like these, you can help organize and consolidate your learning if you can answer these questions in a well-organized and complete manner.

Do not think that it is easier to write adequate answer for brief essays than for longer essays. Short-answer essays may be more challenging because answers are required to be both brief and complete. And after you have answered these questions below, construct some similar questions of your own devising, and then answer them. The Study Questions at the end of each textbook chapter make very good short-answer essay questions.

1. Summarize the history of remarriage trends in the U.S., as set forth in the textbook.

2. What are the differences between remarriage in which only one person has been married before, as compared to remarriages, in which both persons have been married before?

3. Ex-husbands more than ex-wives want to remarry, and do. Why?

ESSAY QUESTIONS

The following are sample essay questions — questions of the type you may be asked if your instructor uses essay questions. Even if your instructor does not use essay questions, you can help organize and consolidate your learning if you can answer these questions in a well-organized and complete manner. Of these essay questions, the third essay question is usually the most challenging.

1. Explore the extent to which remarriages are happy and stable. When they are, why are they? When they are not, why are they not?

2. Remarried couples frequently state that it seems difficult for them to "fit in" and that they often don't know what's expected of them. According to the text, why is this?

3. Imagine that a person you know is getting remarried and wants to know what to expect. At lunch with this friend, you explain the things you think are most important to know about remarriages. What do you tell our friend? Give a well-organized answer, taking care to be specific and complete.

ANSWERS

CHAPTER SUMMARY

after ***widowhood***
homogamy tends
a cultural ***script***
the ***stepparent***
role ***ambiguity***

COMPLETION

1. stepmother trip
2. hidden agenda
3. quasi-kin
4. double remarriage
5. remarried families
6. single remarriage
7. double remarriage

TRUE-FALSE

1. T
2. T
3. F
4. T
5. T
6. F
7. T
8. T
9. F
10. F
11. F
12. T
13. F
14. F

MULTIPLE CHOICE

1. D
2. A
3. C
4. D
5. A
6. D
7. A
8. D
9. A
10. D
11. D